Pseudo-Populations

Andreas Quatember

Pseudo-Populations

A Basic Concept in Statistical Surveys

 Springer

Andreas Quatember
Johannes Kepler University Linz
Department of Applied Statistics
Linz, Austria

ISBN 978-3-319-35280-0   ISBN 978-3-319-11785-0 (eBook)
DOI 10.1007/978-3-319-11785-0

Springer Cham Heidelberg New York Dordrecht London
© Springer International Publishing Switzerland 2015
Softcover re-print of the Hardcover 1st edition 2015

Printed on acid-free paper

Springer International Publishing AG Switzerland is part of Springer Science+Business Media
(www.springer.com)

Preface

My first encounter with the term "pseudo-population" was when I was involved in the DACSEIS project of the European Union (Data Quality in Complex Surveys within the New European Information Society), wherein I was responsible for the Austrian contribution. In this project, among other things, artificial or "pseudo-populations," herein called "plausible or synthetic universes," generated on the basis of data from the relevant national Labour Force Surveys of the participating countries, were used to analyze the quality of different methods to estimate the variance of labor force statistics such as the number of employed or unemployed people by simulations (cf. Münnich et al. 2003).

Moreover, while teaching students of the bachelor's and master's programs in statistics at my university in courses on sampling methods, I realized that the picture of generating a pseudo-population has the potential to substantially improve the students' understanding of various concepts in sampling theory and survey methodology. One can describe, for instance, the Horvitz–Thompson estimator of the total of a study variable, a statistic fundamental in sampling theory, by the generation of a pseudo-population estimating the original finite population with respect to this parameter. For this purpose, the variable values observed in the sample are assigned to the units of the pseudo-population by replicating each of these values by a factor that reflects both the sampling and estimation process. As a rule, such replication factors are not integers. The Horvitz–Thompson estimator of the total in the original population then is nothing else but the total of the same variable in the pseudo-population consisting of not only whole units, but also parts of whole units. Further concepts of estimators of the total, such as the ratio or regression estimator, can be illustrated in the same way. The difference in these concepts is explained by a different composition of the pseudo-population with respect to the study variable.

The point is that the pseudo-population concept unites several features of statistical sampling theory such as sampling techniques and estimation methods under a single roof. In my experience, this approach may really help to communicate a basic understanding of such techniques, by the formulae often not intuitively comprehensible for all students, especially those with little knowledge of the probability theory. With this fundamental methodological understanding, students and other users should be able to focus on questions concerning the difference between methods and their practical implementation in statistical software such as R (for recent books see, for instance, Lumley 2010; Kauermann and Küchenhoff 2011).

Therefore, the first part of this monograph can be seen as a textbook using the unified approach of the pseudo-population concept to describe different aspects of the sampling theory and survey methodology. After an introductory chapter on statistical surveys, in Chap. 2, the pseudo-population concept is introduced to describe the rationale behind the Horvitz–Thompson estimator. Subsequently, the concept is applied to the presentation of different probability and non-probability sampling methods and the alternative estimation procedures of ratio estimation, regression estimation, and iterative proportional fitting using auxiliary information. This chapter is continued by the presentation of the estimation of special totals such as the size of the original population by the capture–recapture method and the cumulative distribution function of a variable under study. It is complimented by other examples of the application of the generation of pseudo-populations in statistical surveys such as the estimation of covariance, measures of associations, small area estimation, and two-phase sampling.

Chapter 3 is concerned with the practical problems of "Nonresponse and Untruthful Answering." The statistical methods of weighting adjustment and data imputation compensating for nonresponse are also described intuitively within the presented pseudo-populations approach that runs like a thread through this work.

The second part of this monograph is concerned with the fields of sampling theory and survey methodology, in which the author applied the pseudo-population concept in his research. In Chap. 4, the basic principles of simulation studies in the area of survey sampling are presented. The basis for such studies is formed by either the original population or a close-to-reality substitute—another pseudo-population. An example of its usage in a secondary analysis of the Austrian PISA (Programme for International Student Assessment) data concludes this chapter (cf. Quatember and Bauer 2012).

Actually, the term pseudo-population was first used in the relevant literature on a certain type of the bootstrap method for a finite population, which is used as a computer-intensive alternative to estimate the sampling distribution of sample statistics (see Chap. 5). In this context, the generation of a pseudo-population, also called the "bootstrap population," is the necessary step between the original sample already drawn and the resamples needed to mimic the interesting sampling distribution. The most important question for researchers to answer for this bootstrap procedure is how to create a bootstrap population, which may serve as an adequate basis for the resampling process with respect to the statistical properties of the estimator under investigation. A review of approaches to answer this fundamental question is presented in Sect. 5.2. Moreover, a direct implementation of the Horvitz–Thompson approach of generating a pseudo-population allowing not only for whole units in the bootstrap population is discussed in some detail as an application of the concept to this resampling technique (cf. Quatember 2014b). It can be seen as a natural complement of recently published works in this field. An example of the application of the finite population bootstrap concludes this chapter.

In Chap. 6, the problem of nonresponse and untruthful answering is addressed again, when the concept of pseudo-populations is applied to "Generalized Randomized Response Questioning Designs," which are privacy protecting alternatives to direct questioning. In this context, this concept serves as the basis to extend theory presented so far mostly for simple random sampling with replacement to general probability sampling schemes. This is of great importance because in the fields where these methods are used such as empirical social or health research, more often than not complex probability sampling schemes are used. Quatember (2009, 2012) presented generalizations of specific families of such strategies applicable to binary variables. Herein, such standardizations are presented for general categorical as well as quantitative variables. The presented frameworks encompass strategies already published and, at the same time, all other combinations of the questions and instructions included in these families not yet published in individual articles. Such methods are currently used in many occasions (cf. Lensvelt-Mulders et al. 2005), but not really present in opinion and market research, where, if it would be possible (or necessary) to leave the usual paths, it has the potential to raise the quality of surveys on sensitive attributes such as in opinion polls.

The concluding chapter on "A Unified Framework for Statistical Disclosure Control" (Chap. 7) considers statistical techniques, which are, without doubt, different from the methods discussed in the preceding chapters because the aim is not to improve the efficiency of survey results. On the contrary, the term statistical disclosure control summarizes methods, which reduce data quality in a controlled way to increase data privacy. This field has become increasingly important in recent years. In the summer of 2013, for instance, according to an article published in the German news magazine *Spiegel*, data privacy activists reported that the computer center of the pharmacists in South Germany sold insufficiently encrypted patient data to market research companies. Basically, trading with such data is legal when data are handed out and used in an encrypted way. These microdata files contain information that may be used in market research of pharmaceutical companies. In this particular case, the identity of patients is only veiled by a code, which enables the known user to conclude on the actual security numbers of the patients. In addition, the age and sex of patients are included in the file. Hence, pharmaceutical companies could possibly understand which medical practices prescribe which medication. Data privacy activists point out that in this way, the companies could be enabled to control the success of the work of their sales staff. A similar problem was discussed at the same time in Austria, where doctors directly sold patients' data.

In this chapter, the concept of pseudo-populations is applied to a group of disclosure techniques to create a comprehensible unified framework for this "CSI family" of methods (cf. Quatember and Hausner 2013). The abbreviation CSI stands for a process that starts with the cloning of variables, followed by the artificial suppression of the values of these clones and is finished with an imputation for these suppressed values. At the end of the day, the original variable is deleted from the file and its masked version has to take over its tasks. In this context, the description of the estimation process by the idea of generating a pseudo-population marks the very moment when the balancing act between the mandatory data protection and the understandable demand of third parties for access to survey data takes place.

Since this book contains my teaching as well as research experiences in the field of data quality in statistical surveys under the unified approach of the pseudo-population concept, I wish to thank all the people who have given me the opportunity to have all these experiences at the Johannes Kepler University, Linz, Austria. In particular, I am grateful to all my colleagues at the Department of Applied Statistics for inspiring me over the years. Especially, I want to thank Werner Müller, head of Data Acquisition and Data Quality, for his continuous support and encouragement. Moreover, I do not want to forget to mention the group of people, who may not really realize that they are my best motivator year by year—the students.

Linz, Austria   Andreas Quatember
June 1, 2015

Contents

Chapter 1
Statistical Surveys

1.1 Introduction

Looking at our everyday life, we find that nearly everything that we see, taste, hear, smell, or touch is just a part of a whole. The unconscious conclusion from such information on the specific "reality" incompletely described by these observations has probably always been part of (not only) human behavior. This pattern was originally developed as a survival strategy passed on from one generation to the other. It ensured the survival of the clan (or the pack) through the evaluation of signals with respect to potential food or imminent danger.

In contrast to such an unconscious behavior pattern, a conscious conclusion on the whole based on a part is seemingly specific to the human race. Each one of us applies this technique, for instance, every time when we cook. A little taste of the dish stands for the quality of the whole food. Wine tasting follows the same inference principle: A small sip poured from any barrel stands for the whole barrel or even for the whole year. The same strategy is used when we try a new fragrance before buying a regular-sized bottle in a perfumery or when we try the sweetness of strawberries before buying the whole quantity needed from a market stand. Moreover, the same principle applies when we read the individual costumer reviews or the average rating for a product offered on online shops such as Amazon or iTunes. Also, when taking blood tests or exams, everybody of use gained practical experience in the application of this method. In all these cases, we are apparently convinced that we can infer something about the whole from observations of a part.

The statistical method of sampling data had its first application in official statistics as a method to estimate the size of populations (for the history of sampling, see, for example, Bethlehem 2009). Sampling theory itself, meaning the theoretical foundation of the conclusion from sampled data to parameters of interest, has been developed only since the end of the nineteenth century (cf., for instance, Kruskal and Mosteller 1980, or Bellhouse 1988). One can be convinced of the resounding success of sampling theory day by day, when listening to any news show on radio or television or when reading newsletters or magazines where the outcomes of market

© Springer International Publishing Switzerland 2015
A. Quatember, *Pseudo-Populations*, DOI 10.1007/978-3-319-11785-0_1

or opinion research and academic empirical studies are presented. It is no surprise that in our knowledge-based society, there is a constantly increasing demand for objective information about issues of interest. Therefore, the application of such a theory has become quite indispensable. Two important application examples are the European Labour Force Survey conducted by the National Statistical Offices, which has an important function in developing the labour market policy of individual nations as well as of the whole European Union, and the international PISA test (Programme for International Student Assessment) of the OECD (Organization for Economic Cooperation and Development), with its function in the evaluation and comparison of educational policies and their effects (for an overview of both surveys, see Eurostat 2012 and OECD 2012). Oftentimes, the media response to the results of surveys like those previously mentioned completely ignores the obvious fact that such surveys are sample, not population, surveys. As such, their results are exposed to natural sample-to-sample fluctuations. This is precisely where the task of statistical sampling theory comes in: the accuracy of survey results and how this accuracy can be influenced and controlled.

1.2   High-Quality Surveys

We define a statistical survey as a survey conducted with the aim of obtaining information from a finite set of survey units, called the population, concerning the frequency distributions of variables of interest or parameters characterizing these distributions. Examples of such a population are the households in a region, the pupils of a certain age class, or the population of eligible voters. Variables of interest are, for instance, the consumption expenditure of households, the employment status of household members, the scores of students in competence tests, or voting behavior. Parameters of interest include the average consumption expenditure per household, the unemployment rate, the average score of students' performances, the recent proportions of different political parties, or any other statistical indicator characterizing the relevant population. A statistical survey can be conducted in the whole population or in just a part of it, called a sample.

The preconditions for a good quality of statistical survey with respect to unknown population distributions or parameters can be summarized under the term representativeness. The representativeness of a survey can be defined in the following way (cf., for instance, Quatember 1996a, or Gabler and Quatember 2012): A survey is called "exactly representative" with respect to an interesting distribution or parameter, if this distribution or parameter is exactly reproduced in the survey (compare with the term "balanced sampling design" in Deville and Tillé 2004, p. 895). It is called "representative," if the distribution or parameter can be estimated (approximately) unbiasedly and the estimation meets a previously defined accuracy requirement. Finally, a survey is "not representative" at all if it is neither exactly representative nor representative.

In this definition, the term "exact representativeness of a survey" corresponds to the optimal target. In the practice of surveys, this meets a rather procedural descriptive purpose, as can be seen in Sects. 2.4.1, 2.4.4 and 2.5. Furthermore, the representativeness of a survey is described, first, by the statistical similarity concept of unbiasedness (see Sect. 2.1) and, second, by a requirement with respect to the efficiency of the estimator. Following this definition, the representativeness of a statistical survey is a precondition for high-quality inference from sample to population on a probability basis. It implicitly includes

- the use of a selection procedure, the sampling technique, to select the survey units of the population for the sample, which allows such a conclusion;
- the application of an appropriate estimation method;
- the choice of a large-enough sample size to meet the requirements for the efficiency of the estimate for a given sampling technique and estimation method; and
- the avoidance or at least the consideration of errors that cannot be explained by carrying out the survey in a sample and not in the population

as necessary conditions (cf. Gabler and Quatember 2013).

1.3 Sampling and Non-sampling Error

The first three items at the end of the previous section refer to the so-called sampling error. This term describes the sample-to-sample variation of estimators occurring through the collection of data in a sample and not in the entire population. The magnitude of the sampling error is determined by all three items, which describe the whole sample design. The last of the four items refers to the non-sampling error that could occur in sample as well as population surveys. There are various possible sources for this category of errors in surveys (see, for instance, Groves et al. 2004, Chaps. 7–10). The sampling frame is an available list of potential survey units, which can serve as a basis for their selection. Its quality depends on the relation between the true population of interest and the available frame population. In this context, under-coverage occurs if units belonging to the true population of interest are not listed in the frame population. On the other hand, we speak of over-coverage if units that are not part of the target population are included in the sampling frame. It is quite clear that in such situations, estimators can only relate to parameters or distributions of the frame population and not the actual population under study. Their presence might be ignored if both types of coverage errors are negligible. With non-negligible frame imperfections, estimation techniques have to be adapted. For over-coverage, the theory of domain estimation might offer an appropriate approach (cf., for instance, Särndal et al. 1992, Chap. 10). For under-coverage, ratio estimation may help (see Sect. 2.6.1). Also other methods of defining the population may be applied. Such alternatives include dividing the population into small areas (of a region, for instance) covering the whole population instead of using a population register or generating telephone numbers instead of using a telephone register (see, for instance, Gabler and Häder 2009).

The term "nonresponse" refers to the "missing data problem," which occurs when a selected survey unit cannot be contacted, completely refuses to participate, or provides no response for at least one of the survey variables. Of course, this is a potential source of non-sampling errors. Therefore, everything should be done to avoid the occurrence of nonresponse (for different approaches see Sects. 3.1 and 6). If the amount of such occurrence is not negligible, so that the statistical analysis cannot be based on the "available cases," nonresponse has to be compensated on the basis of model assumptions applying the methods of weighting adjustment and data imputation as relevant statistical strategies (see Sects. 3.2 and 3.3).

With regard to the estimation quality, wrong answers are even worse than nonresponse because they are usually indistinguishable from true answers. Strategies to reduce untruthful answering have been developed, for instance, in the field of empirical social research (see again Sect. 3.1). Also, the statistical methods of randomized response address the problems arising from nonresponse and untruthful answering (see Sect. 6). These are probabilistic-driven alternatives to direct questioning on sensitive issues such as drug use, domestic violence, harassment at work, or voting behavior. In all these alternatives, the actual question answered stays unknown to the data collector. The idea is that not enabling the data collector to link the given answer to the sensitive variable should increase the respondent's willingness to cooperate.

Other potential sources of non-sampling errors are a poor questionnaire (cf., for instance, Groves et al. 2004, Chaps. 7 and 8), the performance of the interviewer (cf. Groves et al. 2004, Chap. 9), and sloppy data entry in the statistical software used (cf. Groves et al. 2004, Chap. 10).

All of the items discussed (and many more not mentioned here) affect the representativeness of statistical surveys and have to be taken into account in order to manage a representative, and not just informative, statistical survey, from which it should be possible to draw probabilistic-based conclusions about a population.

Chapter 2
The Pseudo-Population Concept

2.1 The Formulation of the Problem

Classical sampling theory addresses the effect of different sampling designs consisting of a sampling method and estimation technique, on the efficiency of the estimation of a parameter under study. Note that sampling design is used with different meanings in the literature (cf., for instance, Särndal et al. 1992, p. 27). In the practice of statistical surveys, totals and functions of totals, such as means, proportions, variances, covariances, correlations, or regression coefficients, cover a large majority of the interesting parameters. Hence, sampling theory traditionally focuses mainly on the estimation of such parameters (cf., for instance, Cochran 1977).

Let U denote the interesting finite population of survey units. This set U consists of N elements characterized by consecutive integers: $U = \{1, \ldots, N\}$ (here and in the following, the notations largely follow Särndal et al. 1992 and Lohr 2010). Let y denote a variable under study and y_k be the fixed value of y assigned to population unit k ($k \in U$): $U \rightarrow \{y_1, \ldots, y_N\}$. The parameter θ of interest may be, for instance, the population total

$$t = \sum_U y_k \qquad (2.1)$$

($\sum_U y_k$ is an abbreviated notation for a sum over all units $k \in U$) or a function of one or more totals. Such is, for instance, the mean value

$$\bar{y} = \frac{t}{N} \qquad (2.2)$$

© Springer International Publishing Switzerland 2015
A. Quatember, *Pseudo-Populations*, DOI 10.1007/978-3-319-11785-0_2

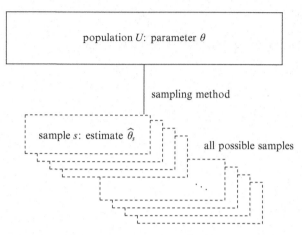

Fig. 2.1 The sampling distribution of a certain estimator $\hat\theta$ within all possible samples s drawn by the same sampling method

or, in the case of a two-dimensional variable (y,x) with assignment $U \to \{(y_1,x_1),\ldots,(y_N,x_N)\}$, the ratio

$$R = \frac{t}{t^{(x)}} \tag{2.3}$$

of the totals t and $t^{(x)} = \sum_U x_k$ of the two variables.

With no census data on the variables of interest available, these population parameters can be estimated by observing the variables under study and possible auxiliary variables, if needed within the estimation process, in a sample $s = \{1,\ldots,n\}$ of n not necessarily distinct elements selected from U by a certain sampling method. For y, this results in the assignment $s \to \{y_1,\ldots,y_n\}$. Note that k and y_k denote the kth element and its value of y either in the population or in the sample depending on the particular context. The observations on the survey variables in the sample are used to calculate the estimates. A census is the special case of a sample survey with $s = U$.

The sample selection follows a certain selection scheme, the sampling method. Such a procedure assigns a certain selection probability to each possible sample s. These probabilities determine the sampling distribution of a certain estimator $\hat\theta$ of parameter θ. For the following, the expected value $E(\hat\theta)$ of $\hat\theta$ and the theoretical variance $V(\hat\theta)$ of this distribution are most important. Both statistical properties provide information on the behavior of the estimator $\hat\theta$ with respect to all possible samples (see Fig. 2.1).

Because the bias of an estimator is defined as the difference between the expected value $E(\hat\theta)$ and the parameter θ, an estimator is called unbiased if $E(\hat\theta) = \theta$ applies. The theoretical variance $V(\hat\theta)$ of an estimator $\hat\theta$ is given by

$$V(\hat\theta) = E[(\hat\theta - E(\hat\theta))^2].$$

Further, its mean square error $\mathrm{MSE}(\hat{\theta})$ is defined by

$$\mathrm{MSE}(\hat{\theta}) = E[(\hat{\theta} - \theta)^2].$$

For an unbiased estimator $\hat{\theta}$, $V(\hat{\theta}) = \mathrm{MSE}(\hat{\theta})$ applies. The random nature of $\hat{\theta}$ is explained by the fact that the sample s, from which the observations on the survey variables are taken to calculate the estimate, is itself a set-valued outcome of a random experiment.

Therefore, to allow design-based conclusions on a population based on the knowledge of the applied sampling design without any modeling, the selection probabilities of all possible samples s of U have to be known in advance. Thus, not any kind of sampling scheme is suitable for such an inference. Sampling methods that allow a design-based inference assign a non-zero inclusion probability to each population unit k ($k \in U$) and a computable probability to each possible sample s. Such methods are called "random" or "probability sampling methods." A sample drawn in such manner is called a "random" or "probability sample" (cf., for instance, Särndal et al. 1992, p. 8).

One way to control the precision of a given estimator is by assigning specific selection probabilities to the units of U. The probability π_k that population unit k will be included in the sample is called the "first-order inclusion probability" of unit k. Furthermore, the probability π_{kl} that two elements k and l will both be included is denoted as the "second-order inclusion probability" of units k and l. For $k = l$, $\pi_{kl} = \pi_k$ applies. These first- and second-order inclusion probabilities are the sums of the selection probabilities of all samples s, which include element k and both elements k and l, respectively.

Let the inclusion of a survey unit k in the sample s be indicated by the sample inclusion indicator

$$I_k = \begin{cases} 1 & \text{if unit } k \in s, \\ 0 & \text{otherwise.} \end{cases}$$

For any probability sampling method, the expectation of the sample inclusion indicator $E(I_k)$ is equal to π_k because $E(I_k) = P(I_k = 1)$ applies. Its variance is given by $\Delta_{kk} \equiv V(I_k) = \pi_k \cdot (1 - \pi_k)$ because $E(I_k^2) = E(I_k)$ applies. The covariance of the two sample inclusion indicators I_k and I_l is given by $\Delta_{kl} \equiv C(I_k, I_l) = \pi_{kl} - \pi_k \cdot \pi_l$ because $E(I_k \cdot I_l) = P(I_k \cdot I_l = 1)$. With $n = \sum_U I_k$, the expectation of size n of a sample for a given sampling method is proven to be $E(n) = \sum_U \pi_k$.

2.2 The Horvitz–Thompson Estimator of a Total

The estimation of the population total t of a variable y is a very important task of statistical sampling theory. In without-replacement sampling, $\sum_s y_k < t$ applies for $s \subset U$. The question arises as to which factors $d_1, \ldots, d_n$ the sampled values

$y_1, \ldots, y_n$ should be "weighted" with to allow an unbiased linear estimator $\sum_s y_k \cdot d_k$ for t when information on auxiliary variables is not available at all (cf. Horvitz and Thompson 1952, p. 667ff). Hence, the following has to apply:

$$E\left(\sum_s y_k \cdot d_k\right) = E\left(\sum_U I_k \cdot y_k \cdot d_k\right) = \sum_U y_k \cdot d_k \cdot E(I_k) = t.$$

Because $E(I_k) = \pi_k$, this equation always holds for $d_k = \frac{1}{\pi_k}$, the reciprocal of the first-order inclusion probabilities ($k = 1, \ldots, N$). Because the π_k-values are determined in the survey's design phase, in which the sampling method is chosen, the multiplicands d_k are called "design weights."

Therefore, the Horvitz–Thompson (HT) estimator of the total t of a variable y,

$$t_{\text{HT}} = \sum_s y_k \cdot \frac{1}{\pi_k}, \tag{2.4}$$

is unbiased and applicable with any probability sample with arbitrary positive first-order inclusion probabilities for all population units k. These design weights are larger for elements having smaller inclusion probabilities, and vice versa. This provides the necessary balance between the inclusion probabilities and the weights given to the observed sample values, which enables the HT estimator to be unbiased for the total of y.

Besides the first-order inclusion probabilities π_k, the probability distribution of all possible samples also determines the second-order inclusion probabilities π_{kl} ($k, l \in U$). These first- and second-order inclusion probabilities are needed to calculate the theoretical variance $V(t_{\text{HT}})$ of the HT estimator, which is given by

$$V(t_{\text{HT}}) = \sum\sum_U \Delta_{kl} \cdot \frac{y_k}{\pi_k} \cdot \frac{y_l}{\pi_l} \tag{2.5}$$

($\pi_k \neq 0 \; \forall \, k \in U$). Provided that $\pi_{kl} > 0 \; \forall \, k, l \in U$,

$$\hat{V}(t_{\text{HT}}) = \sum\sum_s \frac{\Delta_{kl}}{\pi_{kl}} \cdot \frac{y_k}{\pi_k} \cdot \frac{y_l}{\pi_l} \tag{2.6}$$

is an unbiased estimator of the theoretical variance $V(t_{\text{HT}})$. For a sampling method with a fixed sample size n, Eq. (2.5) can be presented as follows:

$$V(t_{\text{HT}}) = -\frac{1}{2} \cdot \sum\sum_U \Delta_{kl} \cdot \left(\frac{y_k}{\pi_k} - \frac{y_l}{\pi_l}\right)^2. \tag{2.7}$$

Under the same conditions as for (2.6),

$$\hat{V}(t_{\text{HT}}) = -\frac{1}{2} \cdot \sum\sum_s \frac{\Delta_{kl}}{\pi_{kl}} \cdot \left(\frac{y_k}{\pi_k} - \frac{y_l}{\pi_l}\right)^2, \tag{2.8}$$

is an unbiased Yates–Grundy–Sen estimator of $V(t_{HT})$ (see Sen 1953; Yates and Grundy 1953).

The quality of the HT estimator t_{HT} does not depend on any modeling. Information can be incorporated in this estimator only by the first- and second-order sample inclusion probabilities in the design phase of the survey, in which the sampling method is determined. Hence, t_{HT} is a pure design-based estimator, meaning that its accuracy depends solely on the applied sampling method, the inclusion probabilities assigned by this method, and the sample size.

2.3 Introduction to the Pseudo-Population Concept

The rationale behind the HT estimation process as expressed by

$$t_{HT} = \sum_s y_k \cdot \frac{1}{\pi_k},$$

in Eq. (2.4) can be described by the idea of generating an artificial population estimating appropriately the original population with respect to the parameter under study. The generation process starts at population U. Each element k of U is assigned a certain value y_k of variable y, but the parameter t is unknown. In the next step, one of all possible samples in Fig. 2.1, which can be drawn according to a given probability sampling scheme (see Sect. 2.4), is realized. In this sample s of n elements, variable y is observed. In the next step, the original population U of size N is estimated with respect to the total t of variable y by a pseudo-population U^*_{HT}. In the concluding step, the HT estimator of t is calculated as the total of the replications of y in U^*_{HT}.

For the generation of the pseudo-population U^*_{HT} (see Fig. 2.2), the variable value y_1 of the first element in the sample is replicated $\frac{1}{\pi_1}$ times delivering $\frac{1}{\pi_1}$ "clones" y^* of y_1 for the pseudo-population; value y_2 of the second sample element is replicated $\frac{1}{\pi_2}$ times delivering $\frac{1}{\pi_2}$ clones, and so on. Hence, the design weights can be seen as the replication factors of this process. Pseudo-population U^*_{HT} has $N^*_{HT} = \sum_s \frac{1}{\pi_k}$ elements. The expectation of size N^*_{HT} is $E(N^*_{HT}) = E\left(\sum_U \frac{1}{\pi_k} \cdot I_k\right) = N$. Its variance depends on the sample size n and the variance of the design weights $\frac{1}{\pi_k}$.

Assuming that all design weights were integers, pseudo-population U^*_{HT} could be characterized by the set $U^*_{HT} = \{1, \ldots, N^*_{HT}\}$ of consecutive integers and the assignment

$$U^*_{HT} \to \left\{ y^*_1, \ldots, y^*_{\frac{1}{\pi_1}}, \ldots, y^*_{\sum_{k<n} \frac{1}{\pi_k}+1}, \ldots, y^*_{N^*_{HT}} \right\}$$

Fig. 2.2 Generating a
pseudo-population for the
Horvitz–Thompson estimator

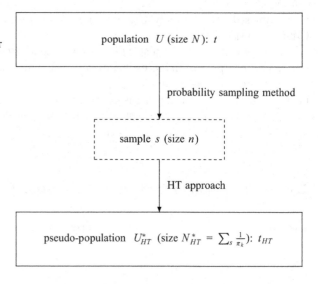

with $y_1^* = \ldots = y_{\frac{1}{\pi_1}}^* = y_1$, $y_{\frac{1}{\pi_1}+1}^* = \ldots = y_{\frac{1}{\pi_1}+\frac{1}{\pi_2}}^* = y_2$, and so on, where
$y_1, y_2, \ldots$ are the values of y in the sample. Hence, the HT estimator (2.4) for
parameter

$$t = \sum_U y_k$$

is presented as

$$t_{\mathrm{HT}} = \sum_{U_{\mathrm{HT}}^*} y_k^*. \tag{2.9}$$

Note that the design weights $\frac{1}{\pi_k}$ are not integers as a rule. Hence, the HT pseudo-
population U_{HT}^* is special in the sense that it may not only contain $\lfloor \frac{1}{\pi_k} \rfloor$ whole units
with the same value y_k of variable y ($\lfloor x \rfloor$ denotes the integer part of $x \in \mathbb{R}$) but also
a $(\frac{1}{\pi_k} - \lfloor \frac{1}{\pi_k} \rfloor)$-piece of unit with that value when $\frac{1}{\pi_k} - \lfloor \frac{1}{\pi_k} \rfloor > 0$ applies ($k \in s$). Also,
for non-integer $\frac{1}{\pi_k}$-values, a presentation of type (2.9) will be used subsequently to
describe the idea of generating a pseudo-population in the estimation process.

Consequently, the efficiency of the unbiased HT estimator t_{HT} for t depends
on the quality of the estimation of U by U_{HT}^* with respect to y or, to be even
more precise, with respect to parameter t. For a census assuming full and truthful
responses, $U_{\mathrm{HT}}^* = U$, $N_{\mathrm{HT}}^* = N$, and $t_{\mathrm{HT}} = \sum_{U_{\mathrm{HT}}^*} y_k^* = \sum_U y_k = t$ applies.

Because the expectation of the pseudo-population size N_{HT}^* equals N but N_{HT}^*
itself does not necessarily have to and, additionally, the design weights $\frac{1}{\pi_k}$ do
not have to be integers, the estimator U_{HT}^* is not the real set-valued Maximum

Likelihood (ML) estimator of population U, which would be the one of all possible populations producing the observed sample with the highest probability (cf., for instance, Casella and Berger 2002, p. 315ff). Nevertheless, U^*_{HT} can be considered as an estimator of U "based on the ML principle." For sampling methods for which N^*_{HT} does not have to equal N, an improvement over the HT estimator may be achieved by correcting the design weights $\frac{1}{\pi_k}$ by the factor $\frac{N}{N^*_{HT}}$. We will return to this idea in Sect. 2.6.1 on the ratio estimator.

In the subsequent section, the pseudo-population concept is applied to different probability sampling methods. In Sect. 2.5, it will help to understand the specific problems of non-probability samples.

2.4 Probability Sampling Methods

2.4.1 Probability Proportional to Size Random Sampling

The unbiased HT estimator t_{HT} of the total t of study variable y (see Sect. 2.2) can be applied to any probability sampling scheme. The most general without-replacement probability sampling method with a fixed sample size to draw the sample directly from the population U is "arbitrary probability random sampling" (π). The π method assigns arbitrary inclusion probabilities $\pi_k > 0$ to all population units k. But, the optimum efficiency of the estimator t_{HT} is ensured by choosing the first-order sample inclusion probabilities $\pi_k \propto y_k$, the "size" of population unit k with respect to y. Under the assumptions $y_k > 0$ and $\pi_k \leq 1 \ \forall \ k \in U$, for $\pi_k = \frac{y_k n}{t}$ applied to all population units, the HT estimator t_{HT} is always equal to t. For fixed n, this can also be seen easily from Eq. (2.7), where with these specific sample inclusion probabilities, the variance $V(t_{HT})$ is zero.

In the HT approach as described by the pseudo-population concept introduced in Sect. 2.3 (Fig. 2.2), the design weights $\frac{1}{\pi_k}$ can be seen as the replication factors of this process. The expectation of the size N^*_{HT} of the generated pseudo-population U^*_{HT} is always N, and its actual value varies depending on the variance of y and the sample size n. In the case of $\pi_k \propto y_k$ with replication factors $\frac{t}{y_k n}$, however, the fact that N^*_{HT} might not be equal to the correct size N of U is irrelevant with regard to the accuracy of the HT estimator. Each generated pseudo-population U^*_{HT} is perfect for the estimation of the interesting total t of y. If it actually contains less than N elements, then this is compensated perfectly by proportionally higher y_k-values in U^*_{HT} than in U, and vice versa. According to the definition in Sect. 1.2, a survey conducted by this sampling mode is exactly representative with regard to the total of y!

Naturally, this idea of choosing the π_k-values in proportion to the y_k-values cannot be translated directly into practice because knowledge of the variable y under study would be needed. Instead, a positive auxiliary variable x known for all population units may serve as a substitute. Herein, this particular sampling method is denoted as "probability π_k proportional to size x_k random sampling" (πPS). For a πPS scheme, the HT estimator t_{HT} according to (2.4) is given by

$$t_{\pi\text{PS}} = \sum_s y_k \cdot \frac{t^{(x)}}{x_k \cdot n}. \tag{2.10}$$

The general π scheme mentioned above is included in πPS sampling by assigning values x_k of an artificial variable x to all population units k in a way that the desired probabilities π_k result from $\pi_k = \frac{x_k n}{t^{(x)}}$. When x is approximately proportional to y, though, choosing the first-order inclusion probabilities according to $\pi_k = \frac{x_k n}{t^{(x)}}$ will consequently result in an estimator with small variance because the survey will be exactly representative with respect to the total $t^{(x)}$. Note that $x_k > 0$ and $\pi_k \leq 1$ must hold true.

For πPS sampling, the HT pseudo-population U^*_{HT} from Sect. 2.3 is denoted as $U^*_{\pi\text{PS}}$. It comprises $\frac{t^{(x)}}{x_1 n}$ replications y^* of sample value y_1, $\frac{t^{(x)}}{x_2 n}$ replications of sample value y_2, and so on. Hence, for the size $N^*_{\pi\text{PS}}$ of $U^*_{\pi\text{PS}}$, $N^*_{\pi\text{PS}} = \sum_s \frac{t^{(x)}}{x_k n}$ applies. The actual value of $N^*_{\pi\text{PS}}$ depends on the values x_k of the sample elements and the sample size n. In the harmonizing pseudo-population concept, $t_{\pi\text{PS}}$ can be presented as

$$t_{\pi\text{PS}} = \sum_{U^*_{\pi\text{PS}}} y^*_k. \tag{2.11}$$

The variance of $t_{\pi\text{PS}}$ does not depend on $N^*_{\pi\text{PS}}$ but on the level of approximate proportionality between x and y. With increasing variance of ratio $\frac{y}{x}$, the variance $V(t_{\pi\text{PS}})$ of $t_{\pi\text{PS}}$ increases. The value of the second-order inclusion probabilities π_{kl} needed for the calculation of (2.5) and (2.6) depends on the practical realization of the πPS scheme (cf., for instance, Kauermann and Küchenhoff 2011, p. 107ff). However, their calculation can be cumbersome for sample sizes occurring in practice (cf., for instance, Särndal et al. 1992, p. 90ff). To be able to estimate the variance of the HT estimator, the much simpler theory of with-replacement proportional to size sampling (see Sect. 2.4.3) might as well be used as its approximation by a computer-intensive method such as the bootstrap (see Chap. 5).

In practice, a πPS sample (or a general π sample) of size n can be drawn, for instance, by applying a specific version of systematic sampling from a list. For this "systematic πPS with random frame order" (Rosén 1997, p. 162), the N population units of the sampling frame have to be randomly ordered. Then, these units are placed end-to-end to each other according to the size of x on a line of length t_x. The first element starts at point zero on the line and has length x_1, the second one ranges from x_1 to $x_1 + x_2$, the third on from $x_1 + x_2$ to $x_1 + x_2 + x_3$, and so on. The sampling interval is calculated by $\frac{t_x}{n}$. In the next step, a random number ξ is drawn from the interval $\left[0; \frac{t_x}{n}\right]$. The first unit selected for the sample s is the unit, which covers value ξ on the line. From there, the other $n - 1$ sampling units are selected by successively adding $\frac{t_x}{n}$ to ξ. Considering the random order of the N population units, this systematic selection mode selects n units from U with first-order inclusion probabilities $\pi_k = \frac{x_k n}{t^{(x)}}$, if only $x_k < \frac{t_x}{n}$ applies for all $k \in U$.

2.4.2 Simple Random Sampling Without Replacement

An example of a probability sampling method, which determines an equal probability for all possible samples by its design, is simple (or unrestricted) random sampling without replacement (SI). The reasons for using this method are that typically the application is rather simple, no auxiliary information is necessary, and the estimation of multivariate relations, for instance, for regression or correlation analysis, is also rather simple compared to other sampling schemes.

For an SI sample, according to the urn model in probability theory of drawing balls without replacement, a fixed number n of all N population units is selected successively without replacement from this population. This can be seen as a special case of πPS sampling with an (artificial) auxiliary variable x having the same value x_k for all population elements resulting in first-order sample inclusion probabilities $\pi_k = \frac{n}{N}$ ($k \in U$) and second-order inclusion probabilities $\pi_{kl} = \frac{n(n-1)}{N(N-1)}$ ($k \neq l$). Therefore, in the HT estimation procedure described by the generation of a pseudo-population estimating the original population (see Sect. 2.3), each sample unit is cloned $\frac{N}{n}$ times to generate the pseudo-population U_{SI}^*, and the representation burden is evenly distributed on all sample elements. Hence, the size N_{SI}^* of U_{SI}^* is always equal to N. In SI sampling, the HT estimator (2.4) for t can be calculated by

$$t_{SI} = \sum_s y_k \cdot \frac{N}{n} = N \cdot \bar{y}_s \qquad (2.12)$$

with $\bar{y}_s = \frac{1}{n} \cdot \sum_s y_k$, the sample mean of y.

In the standardized presentation of the HT approach by the pseudo-population concept, the estimator t_{SI} is presented as

$$t_{SI} = \sum_{U_{SI}^*} y_k^*, \qquad (2.13)$$

which differs from $t_{\pi PS}$ according to (2.11) in the composition of the pseudo-population.

A sample drawn with uniform inclusion probabilities is called a "self-weighting sample." Because each element of a self-weighting sample represents the same numbers of pseudo-population units, the sample measures of various distribution characteristics, such as the mean, any quantile, or the variance, directly estimate the respective population parameters without bias (see also Sect. 2.7).

Derived from (2.7), with the given SI first- and second-order inclusion probabilities the theoretical variance of t_{SI} yields

$$V(t_{SI}) = N^2 \cdot (1 - f) \cdot \frac{S^2}{n} \qquad (2.14)$$

with

$$S^2 = \frac{1}{N-1} \cdot \sum_U (y_k - \bar{y})^2.$$

The "$(N-1)$-variance" S^2 is usually used to harmonize the presentation in sampling theory instead of the "N-variance" S_N^2 because, in simple random sampling without replacement, the sampling variance of y, $S_s^2 = \frac{1}{n-1} \cdot \sum_s (y_k - \bar{y}_s)^2$, is the unbiased estimator of S^2. Hence, an unbiased estimator of $V(t_{SI})$ is given by

$$\hat{V}(t_{SI}) = N^2 \cdot (1-f) \cdot \frac{S_s^2}{n}, \qquad (2.15)$$

and this perfectly harmonizes with the expression of Eq. (2.14).

In practice, also the systematic sampling mode from Sect. 2.4.1 can be applied in order to generate an SI sample. For this purpose, let $x = 1$ for all N units of a randomly ordered population list giving a line of length N. The random number ξ is drawn from the interval $[0; \frac{N}{n}]$. From this starting point all n population elements covering the points $\xi + k \cdot \frac{N}{n}$ on the given line are chosen ($k = 0, 1, \ldots, n-1$).

2.4.3 Random Sampling with Replacement

In contrast to the SI technique, simple random sampling with replacement (SIR) of n elements follows the i.i.d. principles (cf., for instance, Casella and Berger 2002, p. 207). In this case, the first-order inclusion probabilities π_k of population elements k needed in the HT estimator are defined as the probabilities of being included in the sample of $m \leq n$ distinct elements. For general with-replacement sampling, Hansen and Hurwitz (1943) developed an estimator of t using the constant probability π_k' to select element k in the next draw of with-replacement sampling. The Hansen–Hurwitz (HH) estimator is given by

$$t_{HH} = \frac{1}{n} \cdot \sum_s y_k \cdot \frac{1}{\pi_k'}. \qquad (2.16)$$

It is only for $n = 1$ that $\pi_k' = \pi_k$ applies. Also in the HH estimation process, the generation of a pseudo-population, denoted here as U_{HH}^*, as a set-valued estimator of the original population U serves as the basis for the estimation of parameter t (see Fig. 2.3). For U_{HH}^*, the variable value y_1 of the first of n sample elements is replicated $\frac{1}{\pi_1 n}$ times, value y_2 of the second sample element is replicated $\frac{1}{\pi_2 n}$ times, and so on. Assuming that all $\frac{1}{\pi_k n}$-values are integers, pseudo-population U_{HH}^* can

Fig. 2.3 Generating a
pseudo-population for the
Hansen–Hurwitz estimator

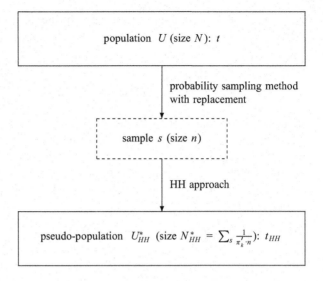

population U (size N): t

probability sampling method
with replacement

sample s (size n)

HH approach

pseudo-population U_{HH}^* (size $N_{HH}^* = \sum_s \frac{1}{\pi_k \cdot n}$): t_{HH}

be characterized by the set $U_{HH}^* = \{1, \ldots, N_{HH}^*\}$ and the assignment

$$U_{HH}^* \rightarrow \left\{ y_1^*, \ldots, y_{\frac{1}{\pi_1 n}}^*, \ldots, y_{\sum_{k<n} \frac{1}{\pi_k n}+1}^*, \ldots, y_{N_{HH}^*}^* \right\}$$

with $y_1^* = \ldots = y_{\frac{1}{\pi_1 n}}^* = y_1, y_{\frac{1}{\pi_1 n}+1}^* = \ldots = y_{\frac{1}{\pi_1 n}+\frac{1}{\pi_2 n}}^* = y_2$, and so on. Hence, the
HH estimator (2.16) for parameter

$$t = \sum_U y_k$$

is expressed by

$$t_{HH} = \sum_{U_{HH}^*} y_k^*. \tag{2.17}$$

Again, in practice, the numbers of replications in the HH process are not integers
as a rule, resulting in a HH pseudo-population including "pieces of units." The size
of U_{HH}^* is $N_{HH}^* = \sum_s \frac{1}{\pi_k n}$.

The theoretical variance $V(t_{HH})$ of the HH estimator is given by

$$V(t_{HH}) = \frac{1}{n} \cdot \sum_U \left(\frac{y_k}{\pi_k'} - t \right)^2 \cdot \pi_k' \tag{2.18}$$

$(\pi'_k \neq 0 \ \forall \ k \in U)$. An unbiased estimator of the theoretical variance $V(t_{HH})$ is given by

$$\hat{V}(t_{HH}) = \frac{1}{n} \cdot \frac{1}{n-1} \cdot \sum_s \left(\frac{y_k}{\pi'_k} - t_{HH} \right)^2 \qquad (2.19)$$

(cf., for instance, Lohr 2010, p. 228ff).

In SIR sampling, $\pi'_k = \frac{1}{N}$ applies for all population units and $N^*_{SIR} = N$, as in the case of SI sampling. However, population elements may be drawn more than once in an SIR sample; due to this information loss, the accuracy of the HH estimator t_{SIR} is lower than that of the HT estimator t_{SI} in finite population surveys:

$$V(t_{SIR}) = N^2 \cdot \frac{S^2_N}{n} \qquad (2.20)$$

with

$$S^2_N = \frac{1}{N} \cdot \sum_U (y_k - \bar{y})^2.$$

An unbiased estimator of $V(t_{SIR})$ is achieved by the estimation of S^2_N by the sampling variance S^2_s, which is an unbiased estimator of S^2_N in SIR sampling.

In the practice of sample surveys, with-replacement sampling schemes are rarely applied. Under certain conditions, however, the sometimes simpler theory of with-replacement techniques may serve as a theoretical approximation in without-replacement procedures, such as for πPS sampling (cf., for instance, Särndal et al. 1992, p. 97ff).

2.4.4 Stratified Random Sampling

Instead of drawing a without-replacement sample unrestricted from the whole population, there may be reasons to split the population with the help of some auxiliary information in different disjoint portions in the survey's design phase. Within each of these subpopulations, any probability sampling method can be applied. The reason for the application of a probability sampling procedure restricted in such a way may be the need for an estimator with given accuracy within so-built subsets of the population or the cost saving operational structure of the survey organization. Most importantly, an appropriate allocation of a given total sample number n on the subsets may increase the efficiency compared to an unrestricted random sample (SI).

Formally, the sampling procedure can be described in the following way: A population U is partitioned into H different subsets U_h $(h = 1,\ldots,H)$ by an any-dimensional categorical auxiliary variable x with H categories. Hence, $U = \{U_1,\ldots,U_H\}$ applies with group sizes $N_1,\ldots,N_H$. The H subsets $U_1,\ldots,U_H$ are called "strata," and variable x is the "stratification variable." For stratified random sampling (ST), a probability sample s_h is selected from each stratum independently from the other strata according to a specific without-replacement probability sampling scheme with element first- and second-order inclusion probabilities π_{hk} for the kth element and π_{hkl} for the kth and the lth elements of stratum h $(k = 1,\ldots,N_h)$. Therefore, the total sample s corresponds to the set union $s = \{s_1,\ldots,s_H\}$. Within each stratum h, the HT estimator

$$t_{h,\mathrm{HT}} = \sum_{s_h} y_k \cdot \frac{1}{\pi_{hk}}$$

of the stratum total t_h is calculated considering the sample inclusion probabilities π_{hk}.

Hence, the HT approach to total estimation (see Fig. 2.2) generates a pseudo-population $U^*_{h,\mathrm{HT}}$ of expected size N_h for the hth stratum by replicating each y-value of sampling units in that stratum π_{hk} times. The estimator of the stratum total t_h is found by summing all the replicated y-values of that stratum in the pseudo-population:

$$t_{h,\mathrm{HT}} = \sum_{U^*_{h,\mathrm{HT}}} y^*_k.$$

Summing all these estimators over all strata gives the HT estimator

$$t_{\mathrm{ST}} = \sum_{h=1}^{H} t_{h,\mathrm{HT}}. \tag{2.21}$$

Therefore, the (imagined) HT pseudo-population $U^*_{\mathrm{ST}} = \{U^*_{1,\mathrm{HT}},\ldots,U^*_{H,\mathrm{HT}}\}$ of the ST sampling scheme consists of H stratum pseudo-populations with expected stratum sizes $N_1,\ldots,N_H$. The expected total size N^*_{ST} of U^*_{ST} is N. Because the strata samples are independent, adding the H theoretical variances of the estimators of the stratum totals according to (2.5) leads to $V(t_{\mathrm{ST}})$, and doing the same with the variance estimators corresponding to (2.6) yields $\hat{V}(t_{\mathrm{ST}})$.

The simplest and most frequent case is stratified simple random sampling (STSI), where simple random sampling of n_h of the N_h stratum elements in the subpopulation U_h is applied within each stratum h $(h = 1,\ldots,H)$. Thus, the first-order inclusion probabilities of elements k belonging to stratum h are given by $\pi_{hk} = \frac{n_h}{N_h}$, and the second-order ones by $\pi_{hkl} = \frac{n_h(n_h-1)}{N_h(N_h-1)}$ uniformly for all $k,l \in U_h$; $t_{h,\mathrm{HT}}$ is actually $t_{h,\mathrm{SI}} = N_h \cdot \bar{y}_{s_h}$ with $\bar{y}_{s_h}$, the sample mean of y in stratum h. Therefore, in the case of STSI, the generated HT pseudo-population U^*_{STSI} always has the correct sizes $N_1,\ldots,N_H$ of all subsets $U_1,\ldots,U_H$ and, consequently, the correct total size N of population U. In other words, a survey conducted by an STSI-sample is, by design, exactly representative with respect to the stratification variable (see Sect. 1.2). For a given total sample number $n = \sum_{h=1}^{H} n_h$ and stratification $U = \{U_1, \ldots, U_H\}$, the HT estimator's overall performance depends on the allocation of n to the H different strata or, equivalently, on how many times the sample elements from different strata are replicated to create U^*_{STSI}. It is the relationship of the different strata replication factors that determines the accuracy of the HT estimator for a given n.

Two special cases of distributing n on the strata in the STSI method are proportional (STSIP) and optimum allocation (STSIO). In the former case, n is allocated in proportion to the relative stratum sizes $\frac{N_h}{N}$ to the strata ($n_h = \frac{N_h}{N} \cdot n$), leading to uniform first-order sample inclusion probabilities $\pi_k = \frac{n}{N}$ for all population units so that all sample elements are replicated $\frac{N}{n}$ times. This results in a self-weighting sample with $t_{\text{STSIP}} = N \cdot \bar{y}_s$. Hence, as soon as the variable under study and the stratification variable are related, for not-too-small populations U and strata U_h, an STSIP sample is more accurate than an SI sample of the same size with regard to the HT estimation of total t. This can be explained by the fact that both the generated pseudo-populations U^*_{STSIP} and U^*_{SI} have the same size $N^*_{\text{STSIP}} = N^*_{\text{SI}} = N$, but at the same number of replications $\frac{N}{n}$ of all units in the samples, the stratum sizes N_h of U are also accurately reproduced in U^*_{STSIP}.

The performance of a sampling design consisting of the estimator $\hat{\theta}$ for parameter θ and a probability sampling method P with expected sample size n is usually compared to the reference sampling design consisting of the same estimator and an SI sample of size n by the ratio

$$d(\hat{\theta}, P) = \frac{V_P(\hat{\theta})}{V_{\text{SI}}(\hat{\theta})}. \tag{2.22}$$

This measure is denoted as the "design effect" of sampling procedure P. In the case of an STSIP sample, for not-too-small populations U and strata U_h, the design effect $d(t_{\text{HT,STSIP}})$ is less than one when not all stratum means are equal.

The optimum "Neyman-Tschuprow allocation" of the total sample size n to the strata with respect to the accuracy of the HT estimator is found by allocating n proportionally to the product of strata sizes N_h and within-stratum standard deviations $S_h = \sqrt{\frac{1}{N_h - 1} \cdot \sum_{U_h} (y_{hk} - \bar{y}_h)^2}$ to the H strata ($h = 1, \ldots, H$). This results in $n_h = \frac{N_h S_h n}{\sum N_h S_h}$ for stratum h (cf., for instance, Särndal et al. 1992, p. 106). Hence, the design weight for element k in stratum h is given by

$$\frac{1}{\pi_{hk}} = \frac{\sum N_h \cdot S_h}{S_h \cdot n}$$

($h = 1, \ldots, H; k = 1, \ldots, N_h$). Strata with higher standard deviations are "overrepresented" in the sample to reduce the total variance of the HT estimator. For the generation of the pseudo-population U^*_{STSIO} of size $N^*_{\text{STSIO}} = N$, this means that the replication factor $\frac{1}{\pi_k}$ is smaller in such strata than in others to compensate for the higher inhomogeneity of the stratum elements with respect to y. This results in a HT estimator t_{STSIO} with even higher efficiency than t_{STSIP} when not all stratum standard deviations of y are equal. For the design effect, $d(t_{\text{HT,STSIO}}) \leq d(t_{\text{HT,STSIP}})$ applies.

Stratifying a population only after the data collection in the estimation phase of the survey may also be useful to increase the efficiency of an estimator (see Sect. 2.6.1). Furthermore, this can be applied to compensate for nonresponse (see Sect. 3.2).

2.4.5 Random Cluster Sampling

The idea of random cluster sampling (C) without replacement is emerged to the costs of face-to-face surveys. For this method, we have to distinguish between selection and survey units. Before the selection process in the design phase of the survey, the population $U = \{1, \dots, N\}$ of survey units is partitioned into, say, M different disjoint subsets $U = \{U_1, \dots, U_M\}$ of sizes $N_1, \dots, N_M$. The population U_{CL} of clusters is numbered as follows: $U_{\text{CL}} = \{1, \dots, M\}$. Such "clusters" of survey elements may, for example, be geographically defined selection units. In a without-replacement cluster sample, from the population of M clusters, m clusters are selected according to any without-replacement probability sampling scheme with first-order sample inclusion probability κ_i for cluster i ($i = 1, \dots, M$). Within the selected clusters, all population units are observed. If not all clusters are of equal size, the sample size n of survey units is a random variable, the actual value of which results from the numbers of population units belonging to the selected clusters: $n = \sum_{s_{\text{CL}}} N_i$. Therein, s_{CL} denotes the sample of m clusters with $s_{\text{CL}} = \{1, \dots, m\}$. Its expectation is $E(n) = \sum_U \pi_k = \sum_{U_{\text{CL}}} N_i \cdot \kappa_i$.

The first-order inclusion probability π_k of a population element k corresponds to the first-order inclusion probability κ_i of the cluster i, of which element k is a member. The interpretation of clusters as survey units and of cluster totals as variable values of these survey units allows the immediate derivation of the HT estimator t_C for the population total t of variable y according to (2.4):

$$t_C = \sum_{s_{\text{CL}}} t_i \cdot \frac{1}{\kappa_i} \tag{2.23}$$

with t_i, the total of y in the ith of the m sample clusters ($i = 1, \dots, m$). In the unified presentation of the HT approach to the estimation of t, for the C method, the HT estimator (2.23) can be presented as

$$t_C = \sum_{U_C^*} y_k^* \tag{2.24}$$

with the replications y^* of the y-values in the sample s of survey units. The pseudo-population U_C^* of survey units generated in the HT procedure of the estimation of t consists of $\frac{1}{\kappa_i}$ replications of each sample cluster i or, in other words, $\frac{1}{\kappa_i}$ clones of all elements belonging to these clusters. The size of U_C^* is $N_C^* = \sum_{s_{CL}} \frac{1}{\kappa_i} \cdot N_i$, which results in N if and only if both the N_i- and the κ_i-values or their product is equal for all survey units.

Using the analogy above, the HT estimator (2.28) might as well be presented as

$$t_C = \sum_{U_{CL}^*} t_i^*$$

with the replications t^* of the cluster totals in the sample s_{CL} of selection units constituting the cluster pseudo-population U_{CL}^*. Moreover, Eqs. (2.5) and (2.6) can also be applied to derive the theoretical variance and an unbiased variance estimator of t_C using κ_{ij}, the second-order cluster inclusion probabilities ($i \neq j; i, j = 1, \dots, M$).

Simple random cluster sampling (SIC) is a special type of cluster sampling, where the clusters are selected according to the SI method. The expected sample size is $n = m \cdot \overline{N}$ with $\overline{N}$, the average size of the clusters in U. For π_k, $\pi_k = \kappa_i = \frac{m}{M}$ with $k \in U_i$ applies for all population elements k ($k \in U$). Thus, an SIC sample is self-weighting at the cluster level, with

$$t_{\text{SIC}} = \frac{M}{m} \cdot \sum_{s_{CL}} t_i. \tag{2.25}$$

The size of the HT pseudo-population $U_C^* = U_{\text{SIC}}^*$, consisting of $\frac{M}{m}$ replications of each y-value from s, varies with the variance of the cluster sizes N_i. For all clusters having the same size, $N_i = \frac{N}{M}$, $N_{\text{SIC}}^* = N$ applies. The second-order inclusion probabilities are given by $\kappa_{ij} = \frac{m(m-1)}{M(M-1)}$. To reduce the variance of the HT estimator in a given SIC sample, ratio estimation can be applied with known population size N to compensate for too-small or too-large pseudo-populations (see Sect. 2.6.1). Compared to an SI sample of the survey units, an SI scheme applied to select clusters of survey units leads, as a rule, to a less efficient estimation $\hat{\theta}$ of a parameter θ at the same expected number of sampling units, meaning that $d(\hat{\theta}, \text{SIC}) > 1$ usually applies. However, it may lead to a more efficient estimation compared to the SI case at the same expected survey costs.

Applying the idea of the πPS technique to cluster sampling, the most efficient way to select the clusters with respect to the accuracy of the HT estimator is to draw them with probabilities $\kappa_i \propto t_i$ ($i = 1, \dots, M$). To be applicable in practice, a good choice for these cluster inclusion probabilities κ_i with unknown t_i-values is to set them proportionally to the known cluster sizes N_i because, more often than not, $N_i \propto t_i$ applies. This sampling technique is called "probability π_k proportional to size of cluster i sampling" (πPSC), which results in the HT estimator

$$t_{\pi\text{PSC}} = \frac{N}{m} \cdot \sum_{s_{CL}} \frac{t_i}{N_i}. \tag{2.26}$$

Within the HT process, for the probabilities $\kappa_i = \frac{N_i m}{N}$, the generated pseudo-population $U_C^* = U_{\pi\text{PSC}}^*$ is always of size $N_{\pi\text{PSC}}^* = \sum_s \frac{1}{\pi_k} = \sum_{sc} \frac{1}{\kappa_i} \cdot N_i = N$. Therefore, $t_{\pi\text{PSC}}$ has a small variance if $N_i \propto t_i$ approximately holds. This high efficiency is paid by more complexity and the variance estimation problem inherent in πPS samples. When $N_i = \frac{M}{m}$ for all clusters of U_{CL}, t_{SIC} is equal to $t_{\pi\text{PSC}}$.

2.4.6 Two-Stage Random Sampling

Within a selected cluster i of a C sample, if not all N_i elements but rather a without-replacement sample of n_i elements belonging to the selected cluster is observed, we speak of "two-stage random sampling" (TS). Any probability sampling method might be applied at both stages of sampling.

TS sampling is the most general of the without-replacement sampling methods. It contains all other without-replacement sampling schemes presented so far as special cases. With πPS sampling at both stages and $M = 1$, the TS sampling scheme comprises the πPS approach to sampling of Sect. 2.4.1 and, therefore, as its special case also the SI sampling mode of Sect. 2.4.2. With $m = M > 1$, it encompasses also the ST method of Sect. 2.4.4. Finally, a TS sample with $m < M > 1$ and $n_i = N_i$ is nothing else than the cluster sample of Sect. 2.4.5. Also complex sampling schemes can be derived from the TS approach to sampling because the clusters of the first stage themselves may be stratified.

For the TS method, the first-order inclusion probability of a population element k belonging to U_i is given by $\pi_k = \kappa_i \cdot \pi_{k|i}$, the product of the selection probability κ_i of selecting cluster i as primary sampling unit and the conditional selection probability $\pi_{k|i}$ of selecting element k as secondary sampling unit, if cluster i is selected ($k = 1, \ldots, N; i = 1, \ldots, M$). For the generation of the HT pseudo-population U_{TS}^* with the TS method, each element k of the sample is replicated $\frac{1}{\kappa_i \cdot \pi_{k|i}}$ times ($k \in U_i$). Considering the cluster structure of the population, the generation process can be described in the following way (see Fig. 2.4): At the first stage, a pseudo-population $U_{i,\text{HT}}^*$ is created within each sample cluster i by replicating each element k of sample s_i within cluster i a number of $\frac{1}{\pi_{k|i}}$ times ($k \in s_i; i = 1, \ldots, m$). At the second generation stage, each of these cluster pseudo-populations is replicated as a whole $\frac{1}{\kappa_i}$ times. This generates the pseudo-population U_{TS}^* of survey units as a set-valued estimator of U with regard to the parameter of interest.

The HT estimator for a TS sample yields

$$t_{\text{TS}} = \sum_s y_k \cdot \frac{1}{\kappa_i \cdot \pi_{k|i}} = \sum_{\text{SCL}} \left(\sum_{s_i} \frac{y_k}{\pi_{k|i}} \right) \cdot \frac{1}{\kappa_i} \tag{2.27}$$

Fig. 2.4 Generating a
pseudo-population for the
Horvitz–Thompson estimator
in TS sampling considering
the two-stage process

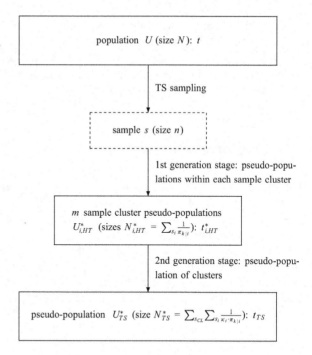

with $\sum_{s_i} \frac{y_k}{\pi_{k|i}}$, the HT estimator of the total t_i of cluster i. In the pseudo-population concept, as simple as always,

$$t_{\mathrm{TS}} = \sum_{U_{TS}^*} y_k^*$$ (2.28)

applies. As usual, y^* denotes the replications of y. For $k \neq l$, the second-order inclusion probabilities for the TS scheme are given by

$$\pi_{kl} = \begin{cases} \kappa_i \cdot \pi_{kl|i} & \text{for } k, l \in U_i, \\ \kappa_{ij} \cdot \pi_{k|i} \cdot \pi_{l|j} & \text{for } k \in U_i \text{ and } l \in U_j \ (i \neq j) \end{cases}$$

(cf., for instance, Särndal et al. 1992, p. 136).

For clusters, homogeneous with respect to the study variable y, one could increase the efficiency compared to t_C by increasing the number of sampled clusters m in the C sample and in turn, to save costs, by taking only a sample of elements instead of all units within each selected cluster.

In two-stage simple random sampling (TSSI), the clusters at the first stage and the elements at the second stage of the sampling procedure are both selected from the respective populations by the SI method. This results in first-order sample inclusion probabilities of $\pi_k = \frac{m}{M} \cdot \frac{n_i}{N_i}$ for unit k in cluster i. Therefore, the size N_{TSSI}^* of HT population U_{TSSI}^* varies with the different cluster sizes N_i. The best performance

Table 2.1 Size N_P^* and replication factors $\frac{1}{\pi_k}$ of the HT pseudo-population U_P^* for without-replacement probability sampling method P

Probability sampling method P	Size N_P^* of pseudo-population U_P^*	Replication factors $\frac{1}{\pi_k}$
πPS	$\frac{t^{(x)}}{n} \cdot \sum_S \frac{1}{x_k}$	$\frac{t^{(x)}}{x_k n}$
SI	N	$\frac{N}{n}$
ST	$\sum_S \frac{1}{\pi_{hk}}$	$\frac{1}{\pi_{hk}}$
STSI	N	$\frac{N_h}{n_h}$
STSIP	N	$\frac{N}{n}$
STSIO	N	$\frac{\sum_{h=1}^{H} N_h \cdot S_h}{S_h n}$
C	$\sum_{S_{CL}} \frac{1}{\kappa_i} \cdot N_i$	$\frac{1}{\kappa_i}$
SIC	$\frac{M}{m} \cdot \sum_{S_{CL}} N_i$	$\frac{M}{m}$
πPSC	N	$\frac{N}{N_i m}$
TS	$\sum_{S_{CL}} \sum_{S_i} \frac{1}{\kappa_i \cdot \pi_{k\|i}} \cdot n_i$	$\frac{1}{\kappa_i \cdot \pi_{k\|i}}$
TSSI	$\frac{M}{m} \cdot \sum_{S_{CL}} N_i$	$\frac{M}{m} \cdot \frac{N_i}{n_i}$

with the TSSI procedure is achieved when clusters of equal size N_i are constructed. The second-order inclusion probabilities π_{kl} in TSSI samples are given by

$$\pi_{kl} = \begin{cases} \frac{m}{M} \cdot \frac{n_i \cdot (n_i - 1)}{N_i \cdot (N_i - 1)} & \text{for } k, l \in U_i \\ \frac{m \cdot (m-1)}{M \cdot (M-1)} \cdot \frac{n_i}{N_i} \cdot \frac{n_j}{N_j} & \text{otherwise.} \end{cases}$$

Eventually, Table 2.1 summarizes the parameters (size and replication factors) of the pseudo-population approach to different sampling techniques.

2.5 Non-probability Sampling Methods

The methods of purposive and arbitrary sample selection are non-probability procedures. Up to a collective failure of these methods in predicting the outcome of the presidential election in the USA in 1948 (cf., for instance, Quatember 2001a, p. 53f), these were the most frequently used sampling methods in opinion and market research. Nowadays, the problems of coverage, the long-term increase in nonresponse rates, and the undeniable benefits of online surveys provide arguments to discuss these methods again (cf., for instance, Baker et al. 2013, p. 6ff). They are in use, for example, "in case-control studies, clinical trials, (or) evaluation research designs" (Baker et al. 2013, p. 1).

Serious examples of purposive sampling are quota, expert choice, and cut-off sampling. These methods orient themselves to certain probability sampling designs. In all these methods, the sample inclusion probabilities of population units cannot be calculated just by the design without assuming a model. Hence, design-based inference cannot be applied.

Quota sampling (QS), a method often applied by of market and opinion research organizations, for example, is oriented towards the STSI sampling scheme. Historically, it was developed in the U.S. market and opinion research field in the 1930s in reaction to the totally arbitrary sample selection modes used until then (cf. Quatember 1996b, p. 29). In the QS scheme, the interviewers' freedom in the selection of sampling units without replacement is limited by known population "quotas" of certain attributes that must be the same in the QS sample. For more than one "quota variable," the quotas can be provided in combined or marginal form (cf., for instance, Quatember 1996b, p. 19ff), resulting in a survey that is exactly representative with respect to the distribution of these combined or marginal variables. On the use of the estimator

$$t_{QS} = N \cdot \bar{y}_s, \tag{2.29}$$

based on the HT estimator in an STSIP sample, Ardilly and Tillé (2006) showed that the extent of its bias depends, on the one hand, on the extent of the deviation of the combined sample quotas from the combined population quotas and, on the other hand, on the covariance of variable y and first-order inclusion probabilities (cf. Ardilly and Tillé 2006, p. 111ff). Apart from rounding errors, the former extent is always zero for quota sampling with combined quotas. Quatember (1996b) showed that this does not apply for QS samples with marginal quotas (cf. Quatember 1996b, p. 64ff). To minimize the impact of the above-mentioned covariance on the bias, the QS sample has to take an STSIP sample with its uniform first-order inclusion probabilities as its model. Various features may help to meet this target: the survey topic itself, the choice of quota variables, the respective interviewer training, the actual interviewer behavior, restrictions with respect to the allowed times and places of interviews, etc. (cf., for instance, Quatember 2001b, p. 96f). In this context, it can be seen that t_{QS} is not a design-based but a model-based estimator of a total t, which only works when the assumed model holds. In practice, this is the crucial point because the adequacy of the model is always disputable.

To estimate t by t_{QS}, a pseudo-population U_{QS}^* is generated presuming uniform first-order inclusion probabilities $\pi_k = \frac{n}{N}$ for all population elements. The quality of U_{QS}^* as an estimator of U with regard to t is bad, if the actual π_k-values are not at least approximately uniform, although the size N_{QS}^* of U_{QS}^* corresponds to the true size N of U. The reason is that the pseudo-population U_{QS}^* is not based on the ML principle, unlike in the HT case, but on a questionable model assumption that will almost never hold in the practice of quota sampling (for the pseudo-population-

Fig. 2.5 Generating a pseudo-population for a model-based estimator in quota sampling

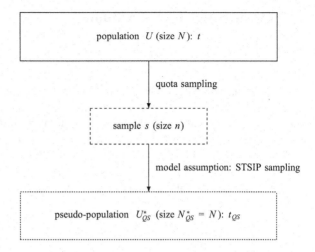

generating algorithm of all purposive and arbitrary sampling schemes exemplarily shown for quota samples, see Fig. 2.5).

Another purposive method is expert choice sampling (EC). This non-probability sampling method is oriented towards the idea of the SIC scheme. "Experts" consciously select one or more clusters of population units that in their opinion are typical for the survey topic. The survey is only investigated within these purposively chosen clusters. This selection scheme can only deliver appropriate estimators for population parameters if the assumed sampling model holds at least approximately. SIC, as well as SI sampling, may be considered for this purpose. With the chosen model, the design-based estimator of the parameter under study is calculated. For the HT estimator, the generated pseudo-population U_{EC}^* can only be appropriate for the estimation of t when the chosen model holds true.

A third example of purposive sample selection schemes is cut-off sampling (CO). This method can be reasonably applied only for the estimation of a highly concentrated total. All the survey units with minor contributions to the sum of y are cut off the population U, and only the few units for which the sum of the y-values almost comes close to the total t are observed. This saves on financial costs at a small reduction in inference quality. However, it must be clear that the estimator t_{CO}, the sum of the observed y-values, always underestimates t. The model applied to treat the subpopulation U_{CO}^* cuts off the original population U because the pseudo-population for U is that of a full survey with respect to the parameter of interest.

Finally, arbitrary sampling does not even try to follow a probability sampling plan. Those who are available are observed. This is not a problem as long as the surveys are purely for entertainment, for instance, of a radio audience. When taking such "convenient" samples in empirical research, for instance, in medicine or psychology to draw statistical conclusions about the characteristics of any population, the underlying model with respect to the sampling procedure must be the SI scheme. With this model, there is a risk that, more often than not, the model-based-generated pseudo-population does not even come close to the actual population regarding the interesting parameters.

2.6   Other Estimators of a Total

2.6.1   The Ratio Estimator

In this section, estimators of the total t of an interesting variable y in the population U under study that "correct" the HT estimator t_{HT} are considered. In contrast to t_{HT} alone, these other estimators can make use of auxiliary information not only in the design phase of the survey but also in the estimation phase after the data collection.

An auxiliary variable is denoted as x. Then, let x_k be the value of x for survey unit k, $t^{(x)}$ its total, and $t_{HT}^{(x)}$ its HT estimator. An example of such an estimator as described above is the ratio estimator t_{rat}, which is calculated by

$$t_{rat} = t_{HT} \cdot \frac{t^{(x)}}{t_{HT}^{(x)}}. \tag{2.30}$$

Therein, t_{HT} denotes the HT estimator of t and

$$t_{HT}^{(x)} = \sum_s x_k \cdot \frac{1}{\pi_k}$$

is the HT estimator for the total of x. The ratio estimator t_{rat} corrects (or "calibrates") the HT estimator t_{HT} with respect to the total $t^{(x)}$ of an auxiliary variable x. The rationale behind this approach is that if in a sample the HT estimator $t_{HT}^{(x)}$ under- or overestimates the known total $t^{(x)}$ of x, then the estimator t_{HT} will very likely too also under- or overestimate the unknown total t of y if y and x correlate positively. Note that for negative correlations of y and x, the so-called product estimator $t_{pro} = t_{HT} \cdot \frac{t_{HT}^{(x)}}{t^{(x)}}$ is preferred (cf., for instance, Cochran 1977, p. 186).

Let us return now to the representation of the concept of the HT estimator by the idea of generating a pseudo-population (see Fig. 2.2). Applying this picture, the reasoning behind the ratio estimator is that a pseudo-population U_{rat}^* instead of the HT pseudo-population U_{HT}^* is generated by the replication of each y- and x-value of s not $\frac{1}{\pi_k}$ times as in the HT approach, but $\left(\frac{1}{\pi_k} \cdot \frac{t^{(x)}}{t_{HT}^{(x)}} \right)$ times. For $t_{HT}^{(x)} \neq t^{(x)}$, the individuals are given higher or lower weights than the design weights to increase or decrease the size $N_{HT}^* = \sum_s \frac{1}{\pi_k}$ of the HT pseudo-population U_{HT}^*. This is done in such a way that the estimator $t_{rat}^{(x)}$ of the total of x is exactly equal to the parameter $t^{(x)}$. Following the notation used in (2.9) and ignoring that the replication factors do not have to be integers, with the clones x^* of the x-values in the sample, $t_{rat}^{(x)}$ can be represented as

$$t_{rat}^{(x)} = \sum_{U_{rat}^*} x_k^* = t^{(x)}.$$

Fig. 2.6 Generating a
pseudo-population for the
ratio estimator

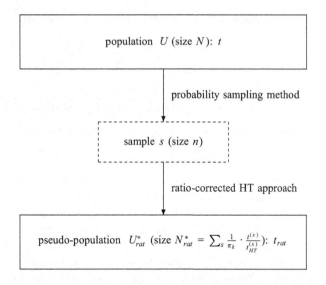

This means that in the light of the sample results, we depart deliberately from
the ML approach in generating the HT pseudo-population. Creating in this way a
pseudo-population U^*_{rat}, the ratio estimator t_{rat} of t is again just the sum over the y^*_k-
values of all (full or piece) units in this "ratio-corrected" pseudo-population U^*_{rat} (see
Fig. 2.6). This completely corresponds to the generation of a typical HT population
U^*_{HT} and the calculation of the estimator t_{rat} using the totals t_{HT} and $t^{(x)}_{HT}$ in U^*_{HT}. This
results in the following alternative expression of the ratio estimator t_{rat} from (2.30):

$$t_{rat} = \sum_{U^*_{rat}} y^*_k = \sum_{U^*_{HT}} y^*_k \cdot \frac{t^{(x)}}{\sum_{U^*_{HT}} x^*_k}. \tag{2.31}$$

The quality of the estimator t_{rat} depends solely on the quality of the generated
pseudo-population U^*_{rat} with respect to the total of variable y, which does not
necessarily mean with respect to population size N.

The appearance of the estimator t_{rat} from Eqs. (2.30) and (2.31) is a result of
the application of a general rule on the estimation of such parameters, which can be
expressed as a function of the population totals of several variables (cf., for instance,
Särndal et al. 1992, p. 162f). The principle is to generate a pseudo-population U^*_{HT}
consisting of the variables involved in the function by replicating the observed
variable values of all of these variables $\frac{1}{\pi_k}$ times ($k \in s$). Then, the unknown totals in
the function are replaced by their HT estimators, which are the pseudo-population
totals. Assuming, for instance, that we want to estimate the ratio

$$R = \frac{t}{t^{(x)}}, \tag{2.32}$$

following this principle, we estimate this function of two population totals by

$$R_{HT} = \frac{t_{HT}}{t_{HT}^{(x)}} = \frac{\sum_{U_{HT}^*} y_k^*}{\sum_{U_{HT}^*} x_k^*}. \tag{2.33}$$

Multiplying R_{HT} by $t^{(x)}$ results in the ratio estimator t_{rat} (2.30). The ratio $\frac{1}{t_{HT}^{(x)}}$ is only approximately unbiased for $\frac{1}{t^{(x)}}$. Consequently, $\hat{R}$ and then also t_{rat} are approximately unbiased estimators for the underlying parameters R and t.

The parameter R is a simple example of a parameter θ, which is a nonlinear function of unknown population totals. Replacing these totals by their HT estimators yields an estimator $\hat{\theta}$ of the nonlinear parameter, for which it is often impossible to calculate an exact expression for its theoretical variance. Approximate expressions for the variance and its estimator can make use of the Taylor linearization technique (cf., for instance, Wolter 2007, p. 230ff). This strategy approximates a nonlinear estimator $\hat{\theta}$ by a proxy estimator, say $\hat{\theta}_T$, which is a linear function of several HT estimators. With this technique, the variance and a variance estimator for this proxy estimator is calculated instead of the exact expressions. The actual substitute $\hat{\theta}_T$ is a result of the first-order Taylor expansion of the nonlinear function $\hat{\theta}$ of estimated totals. In large samples, the linearized estimator $\hat{\theta}_T$ will behave approximately like the original nonlinear estimator $\hat{\theta}$, in which the theoretical variance is needed. As long as the partial derivatives can be calculated, an approximate variance formula and an approximate estimator of the variance can be determined with this technique.

For the Taylor linearization of R_{HT}, for example, the first-order Taylor series expansion R_T of R_{HT} is given by

$$R_{HT} \approx R_T = R + \left[\frac{\delta R_{HT}}{\delta t_{HT}}(t, t^{(x)}) \right] \cdot (t_{HT} - t)$$

$$+ \left[\frac{\delta R_{HT}}{\delta t_{HT}^{(x)}}(t, t^{(x)}) \right] \cdot \left(t_{HT}^{(x)} - t^{(x)} \right)$$

$$= R + \frac{t_{HT}^{(z)}}{t^{(x)}}.$$

Therein, the term $\frac{\delta R_{HT}}{\delta t_{HT}}\left(t, t^{(x)}\right)$ describes the first partial derivative of the function R_{HT} after t_{HT} around the point $(t, t^{(x)})$. Furthermore, $t_{HT}^{(z)}$ denotes the HT estimator of the population total of variable $z = y - R \cdot x$, for which $t_{HT}^{(z)} = t_{HT} - R \cdot t_{HT}^{(x)}$ applies. The point is that the theoretical variance of R_T,

$$V(R_T) = \frac{1}{\left(t^{(x)}\right)^2} \cdot V\left(t_{HT}^{(z)}\right), \tag{2.34}$$

can be used as an approximation of the variance $V(R_{HT})$ of the nonlinear estimator R_{HT}. In (2.34), the term $V(t_{HT}^{(z)})$ denotes the variance of the HT estimator of the population total of variable z,

$$V\left(t_{HT}^{(z)}\right) = \sum\sum_U \Delta_{kl} \cdot \frac{z_k}{\pi_k} \cdot \frac{z_l}{\pi_l}, \tag{2.35}$$

with $z_k = y_k - R \cdot x_k$ $(k \in U)$.

Therefore, the approximate variance (2.34) of R_{HT} can be estimated by

$$\hat{V}(R_{HT}) = \frac{1}{\left(t_{HT}^{(x)}\right)^2} \cdot \sum\sum_s \frac{\Delta_{kl}}{\pi_{kl}} \cdot \frac{\hat{z}_k}{\pi_k} \cdot \frac{\hat{z}_l}{\pi_l}, \tag{2.36}$$

which arises as a result of the substitution of $t^{(x)}$ by $t_{HT}^{(x)}$ and of R by R_{HT}, yielding $\hat{z}_k = y_k - R_{HT} \cdot x_k$ $(k \in s)$.

Returning to the ratio estimator t_{rat} of total t, following (2.30), t_{rat} can be written as

$$t_{rat} = t_{HT} \cdot \frac{t^{(x)}}{t_{HT}^{(x)}} = t^{(x)} \cdot R_{HT}.$$

Hence, its variance results in

$$V(t_{rat}) = \left(t^{(x)}\right)^2 \cdot V(R_{HT}) \approx V\left(t_{HT}^{(z)}\right). \tag{2.37}$$

The theoretical variance $V(R_{HT})$ is approximated by $V(R_T)$ from (2.34). Hence, the variance of the ratio estimator can be approximated by the variance (2.35) of the HT estimator of variable $z = y - R \cdot x$. Following (2.36), $V(t_{rat})$ is estimated by

$$\hat{V}(t_{rat}) = \frac{\left(t^{(x)}\right)^2}{\left(t_{HT}^{(x)}\right)^2} \cdot \sum\sum_s \frac{\Delta_{kl}}{\pi_{kl}} \cdot \frac{\hat{z}_k}{\pi_k} \cdot \frac{\hat{z}_l}{\pi_l}. \tag{2.38}$$

The disadvantage of only approximate unbiasedness of the ratio estimator in comparison with the HT estimator can only be justified if the efficiency of the estimation by t_{rat} increases directly compared to the estimation by t_{HT}. It can be shown that this will be the case when there is a high statistical relationship between study variable y and auxiliary variable x represented by a straight line through the origin of the coordinate system (cf., for instance, Lohr 2010, p. 133).

Ratio estimation and regression estimation (see Sect. 2.6.3) are examples of model-assisted methods. This means that their efficiency compared to the design-based HT estimator depends on the correctness of the underlying model. However, the estimators' approximate unbiasedness and the validity of the variance and

Fig. 2.7 Generating a pseudo-population for the separate ratio estimator in ST sampling

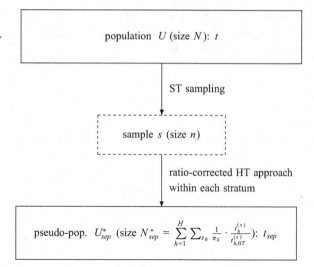

variance estimator formulae are not affected if the model of linear dependency between y and x does not hold.

In ST sampling, the ratio estimator (2.30) combines $\frac{1}{\pi_k} \cdot \frac{t^{(x)}}{t^{(x)}_{HT}}$ replications of each value y_k in s to the overall-pseudo-population U^*_{rat}. In an alternative approach to this combined ratio estimator, ratio estimation is applied in each stratum h to estimate the stratum totals $t_1, \ldots, t_H$ of y ($h = 1, \ldots, H$). This corresponds to the generation of separate pseudo-populations for each stratum. In this case, the replication factor of unit k in stratum h is given by $\frac{1}{\pi_k} \cdot \frac{t^{(x)}_h}{t^{(x)}_{h,HT}}$ ($k \in U_h$). The so-called separate ratio estimator

$$t_{sep} = \sum_{h=1}^{H} t_{h,HT} \cdot \frac{t^{(x)}_h}{t^{(x)}_{h,HT}} \qquad (2.39)$$

may be chosen, if ratios $\frac{t_h}{t^{(x)}_h}$ vary considerably over the strata (see Fig. 2.7).

A special case of ratio estimation is obtained when the size N of population U is used as known auxiliary information. For this purpose, an auxiliary variable x is defined, for which $x_k = 1$ applies for each population unit k. Then, the population total of x is given by $t^{(x)} = N$. Its HT estimator results in $t^{(x)}_{HT} = N^*_{HT} = \sum_s \frac{1}{\pi_k}$, the size of the pseudo-population U^*_{HT}.

With this specific auxiliary variable x, the ratio estimator according to (2.30) results in:

$$t_{rat(N)} = t_{HT} \cdot \frac{t^{(x)}}{t^{(x)}_{HT}} = t_{HT} \cdot \frac{N}{\sum_s \frac{1}{\pi_k}}. \qquad (2.40)$$

If $\sum_s \frac{1}{\pi_k} \neq N$ applies for the sample drawn, t_{HT} is ratio-corrected by $t_{rat(N)}$ in the same proportion that the sum of the design weights $\sum_s \frac{1}{\pi_k}$ is adjusted to N. If $\sum_s \frac{1}{\pi_k} = N$, which is the case in several probability sampling methods (see Table 2.1), $t_{rat(N)} = t_{HT}$ applies.

Looking at the adapted pseudo-population $U^*_{rat(N)}$ (compare with Fig. 2.6), $t_{rat(N)}$ can be described as a HT estimator corrected in the most natural way. It is generated by replicating each y_k exactly $\left(\frac{1}{\pi_k} \cdot \frac{N}{\sum_s \frac{1}{\pi_k}}\right)$ times, ensuring a pseudo-population $U^*_{rat(N)}$ of size $N^*_{rat(N)} = N$, the size of the original population U.

For the nonlinear approximately unbiased estimator $t_{rat(N)}$, approximate expressions for the variance and its estimator can also make use of the Taylor linearization technique. With $R_{HT} = \frac{t_{HT}}{\sum_s \frac{1}{\pi_k}}$ being a natural estimator of the population mean $\bar{y} = \frac{t}{N}$, the ratio estimator $t_{rat(N)}$ can be expressed by $t_{rat(N)} = N \cdot R_{HT}$. For the approximation of variance $V(t_{rat(N)})$, the z-values in (2.35) are given by $z_k = y_k - \bar{y}$ ($k \in U$):

$$V(t_{rat(N)}) \approx \sum\sum_U \Delta_{kl} \cdot \frac{z_k}{\pi_k} \cdot \frac{z_l}{\pi_l}. \tag{2.41}$$

This variance is estimated by

$$\hat{V}(t_{rat(N)}) = \frac{N^2}{(N^*_{HT})^2} \cdot \sum\sum_s \frac{\Delta_{kl}}{\pi_{kl}} \cdot \frac{\hat{z}_k}{\pi_k} \cdot \frac{\hat{z}_l}{\pi_l} \tag{2.42}$$

with $\hat{z}_k = y_k - \frac{t_{HT}}{N^*_{HT}}$ ($k \in s$).

2.6.2   Poststratification and Iterative Proportional Fitting

A type of ratio estimation based on known group sizes can also be used to increase the efficiency of an estimator for t in a sample drawn from a population that was not partitioned into these strata before the sample selection. The corresponding estimation methods are poststratification and iterative proportional fitting (IPF) or raking adjustment (cf., for instance, Bethlehem 2002).

In an observed probability sample s drawn from a population U, which was not stratified by a one- or more-dimensional categorical variable $\mathbf{x}$ in the design phase of the survey, it may appear that the interesting variable y differs in these sample strata. Furthermore, the sizes of the H stratum pseudo-populations built by the HT approach (see Sect. 2.2) may differ widely from the true stratum sizes N_h ($h = 1, \ldots, H$). If both cases apply, we might want to mimic the stratification by x after the data collection.

The poststratified (PST) estimator for t is given by

$$t_{\text{PST}} = \sum_{h=1}^{H} t_{h,\text{rat}(N)} = \sum_{h=1}^{H} t_{h,\text{HT}} \cdot \frac{N_h}{\sum_{s_h} \frac{1}{\pi_k}}. \tag{2.43}$$

This estimator is the sum over the ratio estimators $t_{h,\text{rat}(N)}$ calculated in each poststratum h. Therein, $t_{h,\text{HT}} = \sum_{s_h} y_k \cdot \frac{1}{\pi_k}$ denotes the HT estimator for the total t_h of y in category h of x. The sum $\sum_{s_h}$ denotes the sum over the n_h elements of s belonging to poststratum h, where the sample number n_h is random. The estimator t_{PST} corrects the design weights of the HT estimator (and the replication factors of the generated pseudo-population U_{PST}^*) with respect to the relative category sizes of variable x only after the data collection. As a consequence, the estimated population U_{PST}^* will be correctly distributed over x with the true strata sizes $N_1, \ldots, N_H$ although the sample itself is not exactly representative with respect to this distribution. Therefore, $N_{\text{PST}}^* = N$ also applies.

An example of an application in which poststratification will pay off in terms of accuracy is when an SI sample is drawn. Applying the PST estimator in this case (SIPST) leads to

$$t_{\text{SIPST}} = \sum_{h=1}^{H} \frac{N_h}{n_h} \cdot \sum_{s_h} y_{hk}. \tag{2.44}$$

Hence, instead of U_{SI}^*, a pseudo-population U_{SIPST}^* is generated, for which each sample unit of sample poststratum s_h is replicated not $\frac{N}{n}$ but $\frac{N_h}{n_h}$ times ($h = 1, \ldots, H$). This shall correct for incorrect relative stratum sizes in the HT pseudo-population U_{SI}^*. For large n and not-too-small strata sizes N_h, the effect is that the variance of t_{SIPST} will be close to that of t_{STSIP}, the HT estimator of STSIP sampling (cf., for instance, Lohr 2010, p. 143). This means that the estimation procedure SIPST pays off when $V(t_{\text{STSIP}}) << V(t_{\text{SI}})$ applies.

The same concept is the basis for the IPF (or raking) approach to estimation. IPF may be applied when there is a more-dimensional categorical poststratification variable $\mathbf{x}$, of which in contrast to the requirements for PST estimation only marginal population stratum sizes are known. The process consists of an iterative adjustment of the original HT design weights $\frac{1}{\pi_k}$ of the sampling units k in a probability sample s.

It starts with the first (one- or more-dimensional) marginal poststratification variable by adapting the design weights $d_k = \frac{1}{\pi_k}$ in the same way as in a PST estimator considering only this variable ($k \in s$). As a consequence, the sum of the adapted design weights of the sample elements belonging to the different poststrata of the first variable will equal the true stratum sizes of the first variable in population U. By these first-step adapted design weights, the stratum sizes of all the other marginal poststratification variables are only adjusted to correctly sum to N. In the next step, the new weights are adapted again to sum to the true stratum sizes of the second marginal poststratification variable. This will in turn affect the stratum sizes, calculated by the sum of the current weights over the sample elements in the respective strata, of all the other variables, including the first. The process is continued for all poststratification variables again and again until the stratum sizes of all these variables, calculated from the adjusted weights, deviate from their true marginal distributions in U by no more than a prescribed maximum. With these final weights $d_{k,\mathrm{IPF}}$, the total t of y is estimated by

$$t_{\mathrm{IPF}} = \sum_s y_k \cdot d_{k,\mathrm{IPF}}. \tag{2.45}$$

Assuming full response, the idea of the IPF estimator can be described in the following way: It starts with the generation of a HT pseudo-population U_{HT}^* of size $N_{\mathrm{HT}}^* = \sum_s \frac{1}{\pi_k}$ with the replication factors $\frac{1}{\pi_k}$ for the sample elements k ($k \in s$). In the first iteration step, by an adjustment of the original replication factors, the composition of the HT population is changed to equal the true category sizes of the first marginal poststratification variable and, consequently, the true size N of the original population. In the second iteration step, the pseudo-population generated in the first step of this process is again adjusted by a change of the replication factors calculated for each $k \in s$ in the first step, but this time with respect to the exact representativeness of the second categorical variable. This, in turn, destroys the exact representativeness of the pseudo-population for the first variable. In the next step, the adjustment is done according to the third marginal variable, which again destroys the distribution according to the second variable, and so on. The process is repeated as long as the composition of the pseudo-population with respect to the marginal distributions of the poststratification variables exceeds a prescribed limit of deviation from the true distributions. If the marginal deviations for all categories of all poststratification variables fall below this limit, the process is stopped and the current replication factors are denoted by $d_{k,\mathrm{IPF}}$. When each sample unit k is replicated $d_{k,\mathrm{IPF}}$ times, a pseudo-population U_{IPF}^* of size $N_{\mathrm{IPF}}^* = N$ with the replication variable y^* is generated, in which the IPF estimator is calculated by

$$t_{\mathrm{IPF}} = \sum_{U_{\mathrm{IPF}}^*} y_k^* \tag{2.46}$$

(see Fig. 2.8). Pseudo-population U_{IPF}^* corresponds approximately to the original population U with respect to the marginal distributions of the poststratification variables used. This property did not apply in the HT pseudo-population, but in contrast to the PST estimation, in the IPF estimation this does not apply to the joint distribution of the variables used for poststratification of the population.

Fig. 2.8 Generating a
pseudo-population for the IPF
estimator

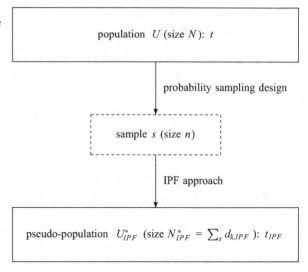

2.6.3   The Regression Estimator

Another example of a model-assisted estimator that uses available auxiliary information in the survey's estimation phase is the regression estimator t_{reg}, of which
the ratio estimator itself is a special case. In contrast to the latter, the regression
estimator can be used without loss of efficiency when the data on y and x in the
scatter plot are based on an arbitrary straight line not necessarily going through the
origin.

Describing the idea for a single auxiliary variable x, known for all population
elements, the regression (reg) estimator develops from

$$t = \sum_U y_k + \sum_U \hat{y}_k - \sum_U \hat{y}_k.$$

For $k \in U$, the proxy y_k-values $\hat{y}_k$ are calculated from the linear regression equation
$y = \beta_1 \cdot x + \beta_2 + \epsilon$ with the residuals $\epsilon = y_k - \hat{y}_k$ (cf., for instance, Särndal et al.
1992, p. 230ff). After some calculation, this yields

$$t = \sum_U \hat{y}_k + \sum_U (y_k - \hat{y}_k)$$
$$= \sum_U (\beta_1 \cdot x_k + \beta_2) + \sum_U (y_k - \beta_1 \cdot x_k) - N \cdot \beta_2.$$

Having information on all x_k-values in the population, the only term that has to
be estimated is the second sum. The HT estimator of the sum $t^{(z)} = \sum_U (y_k - \beta_1 \cdot x_k)$
of variable $z = y - \beta_1 \cdot x$ is given by $t^{(z)}_{\text{HT}} = \sum_s (y_k - \beta_1 \cdot x_k) \cdot \frac{1}{\pi_k}$. Rearranging

Fig. 2.9 Generating a
pseudo-population for the
regression estimator

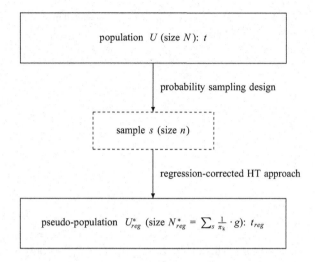

the equation and replacing the regression coefficient β_1 by its estimator b_1 from the
sample (see Sect. 2.8.1) gives the regression estimator

$$t_{\text{reg}} = t_{\text{HT}} + b_1 \cdot \left(t^{(x)} - t_{\text{HT}}^{(x)} \right). \tag{2.47}$$

Equation (2.47) can be presented in the following way:

$$t_{\text{reg}} = t_{\text{HT}} \cdot \left[1 + \frac{b_1}{t_{\text{HT}}} \cdot \left(t^{(x)} - t_{\text{HT}}^{(x)} \right) \right].$$

With $g = 1 + \frac{b_1}{t_{\text{HT}}} \cdot (t^{(x)} - t_{\text{HT}}^{(x)})$, this results in

$$t_{\text{reg}} = t_{\text{HT}} \cdot g \tag{2.48}$$

with $g = 1$ and $t_{\text{reg}} = t_{\text{HT}}$ for $t^{(x)} = t_{\text{HT}}^{(x)}$, which will be approximately fulfilled
for large n. For $t^{(x)} \neq t_{\text{HT}}^{(x)}$, the weight g is not equal to one and t_{HT} is "regression-
corrected" to t_{reg}. For the regression estimator of $t^{(x)}$, $t_{\text{reg}}^{(x)} = t^{(x)}$ applies.

Described by the picture of generating a pseudo-population, the regression
estimation process creates a population U_{reg}^* of size $N_{\text{reg}}^* = \sum_s \frac{1}{\pi_k} \cdot g$. Each sample
value y_k is replicated $\left(\frac{1}{\pi_k} \cdot g \right)$ times, adapting in this way the sum of the replicated
values of auxiliary variable x in the HT pseudo-population U_{HT}^* exactly to $t^{(x)}$ and the
size N_{HT}^* of U_{HT}^* (compare with Fig. 2.2) to N_{reg}^* based on an under- or overestimation
of $t^{(x)}$ by $t_{\text{HT}}^{(x)}$ (Fig. 2.9). This may lead to a more efficient estimation of the original
population U with respect to t.

Table 2.2 Size N^* of pseudo-population U^* generated in different estimation processes for a total t under general π sampling

Estimator	Size N^* of pseudo-population U^*	Replication factors
t_{HT}	$\sum_s \frac{1}{\pi_k}$	$\frac{1}{\pi_k}$
t_{rat}	$\sum_s \frac{1}{\pi_k} \cdot \frac{t^{(x)}}{t^{(x)}_{HT}}$	$\frac{1}{\pi_k} \cdot \frac{t^{(x)}}{t^{(x)}_{HT}}$
$t_{rat(N)}$	N	$\frac{1}{\pi_k} \cdot \frac{N}{\sum_s \frac{1}{\pi_k}}$
t_{reg}	$N^*_{reg} = \sum_s \frac{1}{\pi_k} \cdot g$	$\frac{1}{\pi_k} \cdot g$
t_{greg}	$N^*_{greg} = \sum_s \frac{1}{\pi_k} \cdot g_k$	$\frac{1}{\pi_k} \cdot g_k$

Both the ratio (2.30) and the regression estimator (2.47) consist of the HT estimator and a correction term based on information on x. Both are also approximately unbiased for t. This is true for t_{reg} because β_1 can only be estimated approximately unbiasedly. The theoretical variance of t_{reg} can be approximated through Taylor linearization (see, for instance, Särndal et al. 1992, p. 234ff). For SI sampling, for instance, the variance of the regression estimator $t_{reg,SI}$ is approximately given by

$$V(t_{reg,SI}) = N^2 \cdot (1 - f) \cdot \frac{S^2}{n} \cdot (1 - \rho_{yx})$$

with ρ_{yx} denoting the correlation of y and x in the population.

If a whole auxiliary vector $\mathbf{x}$ consisting of $l \geq 1$ auxiliary variables $x_1, \ldots, x_l$ is available, we can generalize t_{reg} to the general regression estimator t_{greg} using regression theory for the estimation of the regression coefficients β_{1i} (see, for instance, Särndal et al. 1992, p. 225ff). With these estimates b_{1i} $(i = 1, \ldots, l)$, to create the pseudo-population U^*_{greg}, each sample element k is replicated $\left(\frac{1}{\pi_k} \cdot g_k \right)$ times with individual weights g_k depending on the sample s and parameters of the assumed regression model (for a discussion on the role of the model, see, for instance, Särndal et al. 1992, p. 238f). In U^*_{greg}, the pseudo-population totals of the replicated variables x_1^* to x_l^* correspond to the respective population totals of variables x_1 to x_l. For large n, the adjusted replication factors $\frac{1}{\pi_k} \cdot g_k$ will be close to the original design weights $\frac{1}{\pi_k}$.

Table 2.2 provides an overview of the sizes of the pseudo-populations U^* and the replication factors generated in different estimation processes for general π samples (see Sect. 2.4.1).

2.7 The Estimation of Special Totals

Other tasks of statistical surveys concern, for instance, the estimation of the cumulative distribution function of a variable or the estimation of the size of a certain population under investigation. In this section, the concept of generating a pseudo-population is applied to these two tasks.

2.7.1 The Estimation of the Cumulative Distribution Function

The general principle of generating a pseudo-population to estimate an actual population U and calculate estimators of parameters characterizing U by the same characteristics describing the pseudo-population can also be applied for the estimation of the cumulative distribution function or, vice versa, quantiles of the distribution of variable y. The cumulative distribution function $F(y_0)$ of variable y at a certain point $y = y_0$ in the population is defined as the proportion of population units having a variable value $\leq y_0$. Let variable z indicate the possession of the property $y_k \leq y_0$ for element k of U. Then,

$$z_k = \begin{cases} 1 & \text{for } y_k \leq y_0, \\ 0 & \text{otherwise.} \end{cases}$$

applies. Hence, at $y = y_0$,

$$F(y_0) = \frac{t^{(z)}}{N} \tag{2.49}$$

is the population mean of variable z according to (2.2) with $t^{(z)} = \sum_U z_k$, the total number of elements k in U with $y_k \leq y_0$.

The question is, how to estimate $F(y_0)$ in a general without-replacement probability sampling setup with arbitrary inclusion probabilities π_k. This can be done by estimating $t^{(z)}$ and then dividing this estimator by N if the generated pseudo-population contains exactly N units as it would happen with SI sampling. Otherwise, the estimation process may again follow the algorithm in Fig. 2.6 with size N as auxiliary information and $t_{\text{rat}(N)}$ from (2.40) adapted as estimator of $F(y_0)$. Replicating all y_k-values of the sampling units in a selected sample s exactly $\left(\frac{1}{\pi_k} \cdot \frac{N}{N_{\text{HT}}^*} \right)$ times, a set-valued estimator $U_{\text{rat}(N)}^*$ of the original population U is created with respect to y or, more specifically, to $F(y_0)$. Therein, the cumulative distribution function $F(y_0^*)$ of the replication variable y^* is the estimator of $F(y_0)$ in U. Using the replicated z-variable z^* results in the approximately unbiased estimator

$$F_{\text{rat}(N)}(y_0) = \frac{t_{\text{rat}(N)}^{(z)}}{N} = \frac{1}{N} \cdot t_{\text{HT}}^{(z)} \cdot \frac{N}{N_{\text{HT}}^*} = \frac{t_{\text{HT}}^{(z)}}{N_{\text{HT}}^*} = F_{\text{HT}}(y_0). \tag{2.50}$$

Applying the ratio estimator $t_{\text{rat}(N)}^{(z)}$ correcting for estimated sizes $N_{\text{HT}}^* \neq N$ in $U_{\text{rat}(N)}^*$ by a division by N corresponds to the estimation of (2.49) by inserting the HT estimators for both parameters, $t^{(z)}$ and N. Hence, the estimation of the population mean $F(y_0)$ might as well follow the algorithm illustrated in Fig. 2.11. In this case, a HT pseudo-population U_{HT}^* is generated by cloning each sample unit k $\frac{1}{\pi_k}$ times with respect to y. The cumulative distribution function $F_{\text{HT}}(y_0^*)$ of the replication variable y^* in this estimated population U_{HT}^* of size $N_{HT}^* = \sum_s \frac{1}{\pi_k}$ estimates $F(y_0)$ in the same way as $F_{rat(N)}(y_0^*)$ does. Using the replicated z-values z^* from U_{HT}^* results in the following HT based expression of (2.50):

$$F_{HT}(y_0) = \frac{\sum_{U_{HT}^*} z_k^*}{N_{HT}^*}. \tag{2.51}$$

In a self-weighting sample such as an SI or STSIP sample, this estimator of $F(y_0)$ can be calculated directly by the observed sample distribution of y because each sample unit represents the same number of population units.

Consequently, estimates for the median or other quantiles of the population distribution of y can also be calculated directly from the distribution of y^* in the pseudo-population U_{HT}^*. In self-weighting samples, this corresponds to the calculation of these estimators directly in the sample.

2.7.2 The Estimation of the Unknown Size of a Population

In very special cases of statistical surveys, the size N of the population of interest U itself is the parameter to be investigated. Since its development, the capture–recapture (CR) procedure has been used to estimate the size of animal populations such as fishes in a lake (for a historical review, see International Working Group for Disease Monitoring and Forecasting 1995a, p. 1048f). Applications to human populations started with a coverage evaluation in the U.S. census and the estimation of the extent of the registration of human deaths and births (cf., for instance, Sekar and Deming 1949). Other applications of the CR method include physics, empirical social research, and epidemiology (see, for instance, International Working Group for Disease Monitoring and Forecasting 1995b, and the bibliography in Fienberg 1992).

For the purpose of the estimation of N, let y be a variable with unique value $y_k = 1$ for all population units. Then, N can be interpreted as the total of y in U. Hence, in a without-replacement probability sample s drawn from U, the HT estimator N_{HT} of N yields $N_{HT} = \sum_s \frac{1}{\pi_k}$, the number N_{HT}^* of units in the pseudo-population U_{HT}^*. According to (2.9), this can be denoted as

$$N_{HT} = \sum_{U_{HT}^*} 1. \tag{2.52}$$

Obviously, the probabilities π_k have to be known for the calculation of N_{HT}. Hence, the HT estimator of N can only be applied when the calculation of the probabilities π_k does not postulate the knowledge of N or when these probabilities can be determined by a model concerning the actual sampling procedure. A simple random cluster sample, for example, with $\pi_k = \frac{m}{M}$ for $k \in U$ yields $N_{SIC} = \frac{M}{m} \cdot n$, where n is the number of units caught in the cluster sample s_C ($n = \sum_{s_C} N_i$). In such a case, the estimation of N can be done by N_{HT} and without the inclusion of further auxiliary information.

As an alternative to the HT estimation, the above-mentioned CR technique applies, in its basic form, the idea of ratio estimation (see Sect. 2.6.1) to the specific problem of the estimation of N. The auxiliary information needed for ratio estimation is brought into the process before the sample s is drawn. At this pre-sampling stage, a sufficiently large number of C units from U are captured without replacement, marked, and then replaced into the population. Only after that, a probability sample s of size n is selected without replacement from U with arbitrary first-order sample inclusion probabilities π_k for all units $k \in U$. In practice, sometimes, a model concerning the actual sampling procedure must be formulated. Understandably, in wildlife applications of the CR method, SI sampling is the most used sampling model.

Let the auxiliary variable x indicate whether a sample element $k \in s$ is marked ($x_k = 1$) or not ($x_k = 0$). In the sample s, for each element k, variable x is observed. With these observations, the population size N can be estimated by

$$N_{\text{rat(CR)}} = N_{HT} \cdot \frac{t^{(x)}}{t_{HT}^{(x)}} = \sum_s \frac{1}{\pi_k} \cdot \frac{C}{\sum_s x_k \cdot \frac{1}{\pi_k}} \qquad (2.53)$$

(for $\sum_s x_k \neq 0$). The estimator (2.53) corrects the HT estimator N_{HT} of N by adapting the estimated number $\sum_s x_k \cdot \frac{1}{\pi_k}$ of marked sampling units in U to its parameter C. The CR estimator $N_{\text{rat(CR)}}$ can be applied not only if the first-order inclusion probabilities π_k are known or when they can be specified by a corresponding model, but also if these probabilities are equal for all $k \in s$ as it is the case for SI, STSI, or SIC sampling. In such cases, the estimator (2.53) reduces to $N_{\text{rat(CR)}} = \frac{nC}{\sum_s x_k}$ with $\sum_s x_k$ being the number of marked elements recaptured in s.

Looking behind the idea of the estimator (2.53), the estimation of N by the CR method can be described in the following way once again using the picture of the generation of a pseudo-population: In a population U of unknown size N, at the first process stage, C elements are captured and marked, resulting in a "recapture-ready" population U_{CR} of unknown size $N_{CR} = N$. From U_{CR}, a probability sample s is drawn according to any probability sampling method. All sampling units are observed with respect to auxiliary variable x. Using this information, an artificial population $U_{\text{rat(CR)}}^*$ is generated by replicating each element k of s a number of $\left(\frac{1}{\pi_k} \cdot \frac{C}{\sum_s x_k \frac{1}{\pi_k}} \right)$ instead of $\frac{1}{\pi_k}$ times. Counting the number of elements of $U_{\text{rat(CR)}}^*$ leads directly to the estimator $N_{\text{rat(CR)}}$ of N (see Fig. 2.10). Hence, within the generated pseudo-population U_{HT}^*, the estimator (2.53) can be re-written as

$$N_{\text{rat(CR)}} = \sum_{U_{\text{rat(CR)}}^*} 1. \qquad (2.54)$$

Fig. 2.10 Generating a
pseudo-population for the
estimation of population size
using the capture–recapture
method

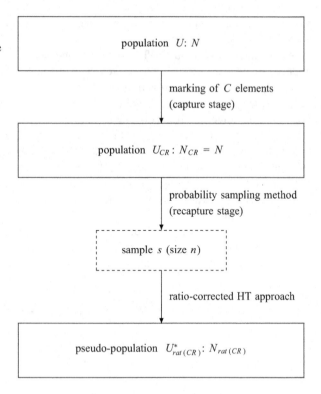

The performance of the estimator (2.53) with different probability sampling
methods may be increased by considering the population structure with respect to
the sampling scheme already at the capture stage of the CR technique. To start with
the most general probability sampling plan with, at most, two stages, to apply TS
sampling (see Sect. 2.4.6), only in each cluster i drawn for the random sample s_C
consisting of m of all M clusters, a predetermined sufficiently large number of C_i
survey units have to be marked ($i = 1, \ldots, m; \sum_{s_C} C_i = C$). Then, after drawing
sufficiently large probability samples of secondary units within the sample clusters,
the cluster sizes N_i of these clusters are estimated by the ratio estimator (2.53)
applied in each of them. Finally, the estimated total size of the sampled clusters
is projected to the number of clusters in the population. The sample inclusion
probabilities needed for the calculation of (2.53) have to be given or modeled. For
random cluster sampling, this approach leads to $N_{rat(CR)} = N_{HT}$.

It follows for the application of the CR technique in general πPS schemes,
which are examples of methods where no such population structure has to be
incorporated that the C population units to be marked are chosen from the entire
population. For $N_{rat(CR)}$ to be calculated with πPS sampling, the first-order sample
inclusion probabilities π_k might only be determined according to an underlying
model concerning the population distribution of the used size variable. For the
application of an ST sampling scheme (see Sect. 2.4.4) with the CR technique, in

its capture phase, a predetermined sufficiently large number of C_h elements have to be marked in each population stratum h to be able to calculate either the combined or separate ratio estimator as presented in Sect. 2.6.1 ($h = 1, \ldots, H$; $\sum C_h = C$). The inclusion probabilities needed can be calculated again according either to the actual sampling scheme or a model of it.

As $N_{\text{rat(CR)}}$ is a ratio estimator, the estimation of N by (2.53) is biased. Furthermore, for small populations and samples, the usual approximate confidence interval calculated with an estimator of the theoretical variance of $N_{\text{rat(CR)}}$ derived from (2.38) is not valid because, in such cases, the distribution of $N_{\text{rat(CR)}}$ may be skewed and hence violate the assumption of being approximately normal. A less biased but still skewed estimator of N was presented for SI sampling by Chapman (1951), and the appropriate variance estimator was presented by Seber (1970). However, an alternative for the construction of valid confidence intervals is provided by the finite population bootstrap method applying the pseudo-population concept (see Sect. 5.4).

2.8 More Estimation Examples

2.8.1 The Estimation of Parameters Such As Covariance

Parameters other than the total of a study variable are often estimated in the practice of statistical surveys, for instance, measures concerned with the relation of variables such as their covariance. The "$(N-1)$-population covariance" $C(y, x)$ of two variables is defined as

$$C(y, x) = \frac{1}{N-1} \cdot \sum_U (y_k - \bar{y}) \cdot (x_k - \bar{x})$$

$$= \frac{1}{N-1} \cdot t^{(y \cdot x)} - \frac{1}{N \cdot (N-1)} \cdot t \cdot t^{(x)}, \quad (2.55)$$

with $t^{(y \cdot x)}$ denoting the population total of the product $y \cdot x$ of the two variables y and x. Furthermore, the "$(N-1)$-variance" $V(y) = C(y, y)$ is consequently defined as

$$V(y) = \frac{1}{N-1} \cdot \sum_U (y_k - \bar{y})^2$$

$$= \frac{1}{N-1} \cdot t^{(y^2)} - \frac{1}{N \cdot (N-1)} \cdot t^2, \quad (2.56)$$

with $t^{(y^2)}$ being the population total of the squared y-values.

To estimate these population parameters, a HT pseudo-population U_{HT}^* is generated with respect to the relevant information. This means that the sample unit k drawn in a probability sample without replacement with first-order inclusion

Fig. 2.11 Generating an HT pseudo-population for the estimation of various parameters of the original population

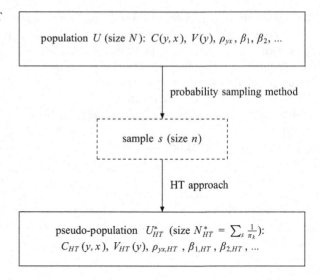

population U (size N): $C(y,x)$, $V(y)$, ρ_{yx}, β_1, β_2, ...

probability sampling method

sample s (size n)

HT approach

pseudo-population U_{HT}^* (size $N_{HT}^* = \sum_s \frac{1}{\pi_k}$):
$C_{HT}(y,x)$, $V_{HT}(y)$, $\rho_{yx,HT}$, $\beta_{1,HT}$, $\beta_{2,HT}$, ...

probabilities π_k is replicated $\frac{1}{\pi_k}$ times regarding the variables y and x under study. This delivers the replicated variables y^* and x^* consisting of $\frac{1}{\pi_1}$ replications of sample values y_1 and x_1, $\frac{1}{\pi_2}$ replications of sample values y_2 and x_2, and so on. Then, the covariance and variance of U are estimated by the covariance and variance in U_{HT}^* (see Fig. 2.11). Note again that, as a rule, the replication factors $\frac{1}{\pi_k}$ are non-integers. Nevertheless, following the general rule of presentation for pseudo-populations consisting of whole and parts of units, as discussed after Eq. (2.9), the estimators of $C(y,x)$ and $V(y)$ can be written as

$$C_{\mathrm{HT}}(y,x) = \frac{1}{N_{\mathrm{HT}}^* - 1} \cdot \sum_{U_{\mathrm{HT}}^*} (y_k^* - \bar{y}^*) \cdot (x_k^* - \bar{x}^*), \qquad (2.57)$$

and

$$V_{\mathrm{HT}}(y) = \frac{1}{N_{\mathrm{HT}}^* - 1} \cdot \sum_{U_{\mathrm{HT}}^*} (y_k^* - \bar{y}^*)^2. \qquad (2.58)$$

The estimators of the relevant parameters in U are nothing else but the same parameters in U_{HT}^*. This is an application of the general rule on the estimation of parameters (cf., for instance, Särndal et al. 1992, p. 162f), which are a function of several population totals, presented in Sect. 2.6.1, for the estimation of a ratio R of two such totals. The estimator (2.57) of $C(y,y)$ can be written as

$$C_{\mathrm{HT}}(y,x) = \frac{1}{N_{\mathrm{HT}}^* - 1} \cdot t_{\mathrm{HT}}^{(y \cdot x)} - \frac{1}{N_{\mathrm{HT}}^* \cdot (N_{\mathrm{HT}}^* - 1)} \cdot t_{\mathrm{HT}} \cdot t_{\mathrm{HT}}^{(x)}, \qquad (2.59)$$

with $N_{\mathrm{HT}}^* = \sum_s \frac{1}{\pi_k}$, the number of units in pseudo-population U_{HT}^* estimating the size N of the original population U, and $t_{\mathrm{HT}}^{(y \cdot x)} = \sum_s y_k \cdot x_k \cdot \frac{1}{\pi_k}$, which can also be written as $t_{\mathrm{HT}}^{(y \cdot x)} = \sum_{U_{\mathrm{HT}}^*} y_k^* \cdot x_k^*$, the sum of the replicated y- and x-values in U_{HT}^*. The use of N_{HT}^*, instead of the possibly known N, follows the considerations explained for the ratio estimator with N as auxiliary information in (2.40). For the estimator (2.58) for $V(y)$,

$$V_{\mathrm{HT}}(y) = \frac{1}{N_{\mathrm{HT}}^* - 1} \cdot t_{\mathrm{HT}}^{(y^2)} - \frac{1}{N_{\mathrm{HT}}^* \cdot (N_{\mathrm{HT}}^* - 1)} \cdot t_{\mathrm{HT}}^2 \tag{2.60}$$

with $t_{\mathrm{HT}}^{(y^2)} = \sum_s y_k^2 \cdot \frac{1}{\pi_k}$ applies. The total of the squared y-values in U is estimated by the sum of the squared y^*-values in U_{HT}^*. Both estimators are not unbiased, but consistent (see, for instance, Särndal et al. 1992, p. 186ff). Because they are nonlinear, Taylor linearization can be used to approximate their variances and estimate these approximate variances (see Sect. 2.6.1).

Moreover, from the artificially generated pseudo-population U_{HT}^*, the estimator $\rho_{yx,\mathrm{HT}}$ of the correlation coefficient ρ_{yx} of y and x in U can directly be calculated with any without-replacement probability sampling scheme by dividing the estimated covariance $C_{\mathrm{HT}}(y, x)$ by the square root of the product of the estimated variances $V_{\mathrm{HT}}(y)$ and $V_{\mathrm{HT}}(x)$ of the two variables y^* and x^* in the HT pseudo-population U_{HT}^*.

The same applies, for instance, to the regression coefficients β_1 and β_2 of a simple regression line $y = \beta_1 \cdot x + \beta_2$ calculated by the least square method. The estimators

$$\beta_{1,\mathrm{HT}} = \frac{C_{\mathrm{HT}}(y, x)}{V_{\mathrm{HT}}(x)} \tag{2.61}$$

and

$$\beta_{2,\mathrm{HT}} = \frac{t_{\mathrm{HT}}}{N_{\mathrm{HT}}^*} - \beta_{1,\mathrm{HT}} \cdot \frac{t_{\mathrm{HT}}^{(x)}}{N_{\mathrm{HT}}^*} \tag{2.62}$$

for β_1 and β_2, respectively, can be calculated directly from the same pseudo-population U_{HT}^* as the coefficients of the regression line in U_{HT}^* (see Fig. 2.11). Both estimators of the regression coefficients are approximately unbiased for the respective parameters. Taylor linearization serves as a tool to develop an approximate variance formula and an estimator of this approximate variance (see, for instance, Särndal et al. 1992, p. 195f).

2.8.2   Measuring Associations in Contingency Tables

The concept of the generation of pseudo-populations in statistical surveys also
provides an instrument to estimate the population association between two categorical variables y and z with r and c categories in a general without-replacement
probability sample s with arbitrary first-order sample inclusion probabilities π_k
($k \in U$). For this purpose, the technique discussed in Sect. 2.7.1 for the estimation
of the cumulative distribution function of a study variable y can be adapted. This
means that in a contingency table with $r \cdot c$ cells, the number N_{ij} of population
units falling into the subpopulation U_{ij} consisting of units belonging to category i of
variable y and category j of variable z can be estimated in the following way using
the HT approach, where $y_k = 1$ applies for all population elements ($i = 1, \ldots, r$,
$j = 1, \ldots, c$):

$$N_{ij,\text{HT}} = \sum_{s_{ij}} \frac{1}{\pi_k}. \tag{2.63}$$

Therein, s_{ij} is the part of the sample s that belongs to U_{ij}. This expression creates
a pseudo-population U^*_{HT} of size $\sum_s \frac{1}{\pi_k}$ consisting of $r \cdot c$ non-overlapping subpopulations $U^*_{ij,\text{HT}}$ ($i = 1, \ldots, r, j = 1, \ldots, c$). The size $N^*_{ij,\text{HT}} = N_{ij,\text{HT}}$ of $U^*_{ij,\text{HT}}$
corresponds to the number $\frac{1}{\pi_k}$ of clones of each element k belonging to s_{ij} ($N^*_{\text{HT}} = \sum_{ij} N^*_{ij,\text{HT}}$).

With known auxiliary information N, according to the special type of the ratio
estimator described in Eq. (2.40) an "N-corrected" estimator of N_{ij} is given by

$$N_{ij,\text{rat}(N)} = N_{ij,\text{HT}} \cdot \frac{N}{\sum_s \frac{1}{\pi_k}}. \tag{2.64}$$

(cf., for instance, Lohr 2010, p. 408). This means that from the sample s, a pseudo-
population $U^*_{\text{rat}(N)}$ is generated, in which the number $N^*_{ij,\text{rat}(N)} = N_{ij,\text{rat}(N)}$ of elements
belonging to subpopulation U^*_{ij} is equal to the sum of the design weight of an
element belonging to s_{ij} multiplied by the "correction factor" $\frac{N}{\sum_s \frac{1}{\pi_k}}$. The total size
of $U^*_{\text{rat}(N)}$ is $\sum_{ij} N^*_{ij,\text{rat}(N)} = N$.

However, the proportion $p_{ij} = \frac{N_{ij}}{N}$ of population units being a member of group
U_{ij} is estimated by

$$p_{ij,\text{HT}} = \frac{N_{ij,\text{HT}}}{N_{\text{HT}}} = \frac{\sum_{s_{ij}} \frac{1}{\pi_k}}{\sum_s \frac{1}{\pi_k}} = \frac{N_{ij,\text{rat}(N)}}{N} = p_{ij,\text{rat}(N)} \tag{2.65}$$

These identical relative category sizes of the combinations of variables y and z in
the pseudo-populations U^*_{HT} and $U^*_{\text{rat}(N)}$ serve as the basis for the calculation of an

estimator of the association of these variables in the finite population U of size N, which is measured, for instance, by Cramér's V with

$$V = \sqrt{\frac{\chi^2}{N \cdot (\min(r, c) - 1)}} \tag{2.66}$$

and

$$\chi^2 = N \cdot \sum_{ij} \frac{\left(p_{ij} - \sum_j p_{ij} \cdot \sum_i p_{ij}\right)^2}{\sum_j p_{ij} \cdot \sum_i p_{ij}}. \tag{2.67}$$

For the estimation of (2.66), within the pseudo-populations U^*_{HT} or $U^*_{\text{rat}(N)}$, the sums

$$\sum_{ij} \frac{\left(p_{ij,\text{HT}} - \sum_j p_{ij,\text{HT}} \cdot \sum_i p_{ij,\text{HT}}\right)^2}{\sum_j p_{ij,\text{HT}} \cdot \sum_i p_{ij,\text{HT}}}$$

or

$$\sum_{ij} \frac{\left(p_{ij,\text{rat}(N)} - \sum_j p_{ij,\text{rat}(N)} \cdot \sum_i p_{ij,\text{rat}(N)}\right)^2}{\sum_j p_{ij,\text{rat}(N)} \cdot \sum_i p_{ij,\text{rat}(N)}}$$

are calculated with $p_{ij,\text{HT}} = p_{ij,\text{rat}(N)}$ estimating p_{ij} (see Fig. 2.12). With the statistic

$$\chi^2_{\text{HT}} = n \cdot \sum_{ij} \frac{\left(p_{ij,\text{HT}} - \sum_j p_{ij,\text{HT}} \cdot \sum_i p_{ij,\text{HT}}\right)^2}{\sum_j p_{ij,\text{HT}} \cdot \sum_i p_{ij,\text{HT}}}, \tag{2.68}$$

Cramér's V (2.66) is estimated by

$$V_{\text{HT}} = \sqrt{\frac{\chi^2_{\text{HT}}}{n \cdot (\min(r, c) - 1)}}. \tag{2.69}$$

Evidently, for statistical tests of independence with general probability sampling, the measure χ^2_{HT} is not distributed as $\chi^2_{(r-1)\cdot(c-1)}$ because the sample counts of the $r \cdot c$ categories are not multinomially distributed. Therefore, other methods for analyzing these associations have been discussed as an alternative to the usual χ^2 test (cf., for instance, Rao and Thomas 1988, p. 235ff, or Thomas et al. 1995).

Fig. 2.12 Generating a
pseudo-population for the
estimation of the association
of two categorical variables
by Cramér's V

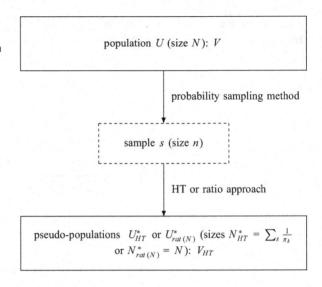

2.8.3   Small Area Estimation

Subgroups U_h of a population U (with $U_h \subseteq U$), for which parameters should be
estimated separately in a statistical survey and which are not defined as explicit
strata in the design-phase of the survey process, are called "domains" or "small
areas." Hence, the number n_h of sample elements in a sample s, which fall in such a
domain, is random. For instance, the total t_h of a variable y within a domain U_h of
size N_h may be estimated unbiasedly by the direct estimator

$$t_{h,\text{dir}} = \sum_{s_h} y_k \cdot \frac{1}{\pi_k} \tag{2.70}$$

(cf., for instance, Rao 2003, p. 15f). This estimator only uses the domain data
set s_h from a given without-replacement probability sample s belonging to the
subpopulation U_h. By (2.70), a pseudo-subpopulation $U^*_{h,\text{dir}}$ is generated as a set-
valued estimator of U_h with respect to t_h by replicating only the n_h elements of s
belonging to domain U_h a number of $\frac{1}{\pi_k}$ times ($k \in s_h$). Hence, the direct estimator
$t_{h,\text{dir}}$ of t_h is a part of the HT estimator t_{HT} of t:

$$t_{\text{HT}} = \sum_{s} y_k \cdot \frac{1}{\pi_k} = t_{h,\text{dir}} + \sum_{s \setminus s_h} y_k \cdot \frac{1}{\pi_k}.$$

Therein, $s \setminus s_h$ denotes the subset of the sample s that does not belong to domain U_h.
In the pseudo-population U^*_{HT} (see Fig. 2.2), with the clones y^* of the sample values
of variable y under study, this expression can be represented by

$$t_{\text{HT}} = \sum_{U^*_{\text{HT}}} y^*_k = \sum_{U^*_{h,\text{dir}}} y^*_k + \sum_{U^*_{\text{HT}} \setminus U^*_{h,\text{dir}}} y^*_k.$$

Fig. 2.13 Generating a
pseudo-population for the
direct estimator of a domain
total

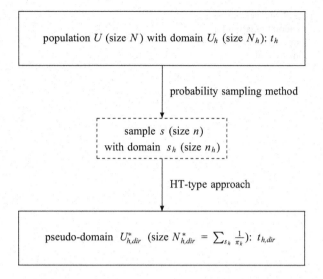

Hence, the estimator $t_{h,\text{dir}}$ of domain total t_h, which is the sum of the y-values of all units belonging to domain U_h of population U, is nothing else but the sum of the y^*-values of all units belonging to "pseudo-domain" $U^*_{h,\text{dir}}$ of pseudo-population U^*_{HT} (see Fig. 2.13). The theoretical variance $V(t_{h,\text{dir}})$ and the variance estimator $\hat{V}(t_{h,\text{dir}})$ are calculated by the usual HT estimator variance and variance estimator summed up only over all survey units belonging to U_h and s_h, respectively (cf., Rao 2003, p. 16).

Looking at the generation process in Fig. 2.13, obviously, an accurate estimate $t_{h,\text{dir}}$ for t_h is possible only for large samples and domains. However, in the case of small domains that are also called "small areas," for the generation of the pseudo-subpopulation $U_{h,\text{dir}}$ as shown in Fig. 2.13, only a small number n_h of sample elements belonging to U_h is replicated. With $n_h = 0$, the domain U_h is not represented in $U^*_{h,\text{dir}}$ at all and $t_{h,\text{dir}}$ cannot be calculated.

An example for a basic model-based estimation technique in the field of small area estimation is the synthetic estimator $t_{h,\text{syn}}$ of the domain total t_h (for details on model-based small area estimation see, for instance, Rao 2003, or Münnich et al. 2004). For the calculation of $t_{h,\text{syn}}$, auxiliary information available for the whole sample can be used. Assuming that x is an auxiliary variable, for which the domain total $t_h^{(x)}$ is known, the synthetic estimator is given by

$$t_{h,\text{syn}} = t_{\text{HT}} \cdot \frac{t_h^{(x)}}{t_{\text{HT}}^{(x)}} \tag{2.71}$$

(cf., for instance, Rao 2003, p. 46ff). The estimators t_{HT} and $t_{\text{HT}}^{(x)}$ in (2.71) are based on data from the whole sample and not just from the part that falls in the domain under investigation. Therefore, under the model that the population ratio $\frac{t}{t^{(x)}}$ is approximately equal to the domain ratio $\frac{t_h}{t_h^{(x)}}$, the estimator $t_{h,\text{syn}}$ performs well.

Fig. 2.14 Generating a pseudo-population for the composite estimator of a domain total

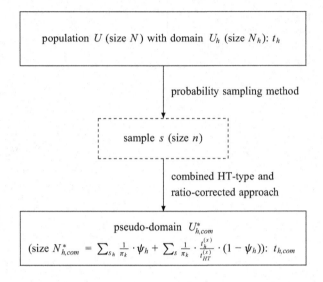

The idea behind this approach is to generate only a pseudo-subpopulation $U^*_{h,\text{syn}}$ as a set-valued estimator of the subpopulation U_h with respect to the estimation of parameter t_h by replicating all the n values of y observed in the sample s $\left(\frac{1}{\pi_k} \cdot \frac{t_h^{(x)}}{t_{\text{HT}}^{(x)}}\right)$ times. In comparison with the generation of $U^*_{h,\text{dir}}$, the pseudo-subpopulation $U^*_{h,\text{syn}}$ is created by the replication of all n sampled units, instead of just the n_h sampling units of sample domain s_h, using smaller expected replication factors as compensation. Based on a larger number of observed elements, its accuracy depends primarily on the validity of the model assumption. If the model concerning the approximate equality of $\frac{t}{t^{(x)}}$ and $\frac{t_h}{t_h^{(x)}}$ does not hold, the synthetic estimator is biased. The greater the deviation of the model and the reality, the larger the bias.

The design-based direct estimator of a domain total is unbiased, but probably inaccurate. The model-based synthetic estimator has a comparatively small variance, but a possible large bias. An estimator that incorporates the advantages of both approaches is the composite estimator (cf., for instance, Rao 2003, p. 57ff), which is defined by

$$t_{h,\text{com}} = t_{h,\text{dir}} \cdot \psi_h + t_{h,\text{syn}} \cdot (1 - \psi_h) \tag{2.72}$$

with $0 \leq \psi_h \leq 1$. The composite estimator generates a pseudo-population $U^*_{h,\text{com}}$ (see Fig. 2.14), which consists of $\sum_{s_h} \frac{1}{\pi_k} \cdot \left(\psi_h + \frac{t_h^{(x)}}{t_{\text{HT}}^{(x)}} \cdot (1 - \psi_h)\right)$ elements that are $\frac{1}{\pi_k} \cdot \left(\psi_h + \frac{t_h^{(x)}}{t_{\text{HT}}^{(x)}} \cdot (1 - \psi_h)\right)$ replications of each unit k from the sample domain s_h and $\sum_{s \backslash s_h} \frac{1}{\pi_k} \cdot \frac{t_h^{(x)}}{t_{\text{HT}}^{(x)}} \cdot (1 - \psi_h)$ elements replicated from the other part $s \backslash s_h$ of the sample s all with replication factors $\frac{1}{\pi_k} \cdot \frac{t_h^{(x)}}{t_{\text{HT}}^{(x)}} \cdot (1 - \psi_h)$. For $\psi_h = 1$, the composite estimator reduces to the direct estimator of t_h with $N^*_{h,\text{com}} = N^*_{h,\text{dir}}$, whereas for $\psi_h = 0$, it reduces to the synthetic estimator with $N^*_{h,\text{com}} = N^*_{h,\text{syn}}$. The amount of replication factors from the two sources depends on the choice of ψ_h. For a good balance between both approaches, with respect to the composition of $U^*_{h,\text{com}}$, the weight ψ_h of $t_{h,\text{dir}}$ should approximate zero when the number n_h of sampling units belonging to domain U_h is small, and vice versa.

2.8.4 Two-Phase Sampling

There are various possibilities to include auxiliary information in the design or estimation phase of a statistical survey. In the absence of such information in the population, the sampling process can be divided into two phases. In phase one, a (large and cheap) without-replacement probability sample s' of n' units is drawn just to observe the auxiliary variable x. In phase two, another without-replacement probability sample s of size $n \leq n'$ is selected as a sub-sample from s' to observe the variable y under study ($s \subseteq s'$). The auxiliary information observed in the first phase can be used at the design and/or the estimation stage of the second phase of the survey (cf., for instance, Lohr 2010, Chap. 12). In both cases, the concept of the generation of pseudo-population can be used to illustrate the estimation process.

An example of an estimator of the total t where auxiliary information from the first phase of the process is incorporated in the design stage of the second phase of a two-phase sampling procedure (tph) is the unbiased HT-type estimator

$$t_{\text{HT,tph}} = \sum_s y_k \cdot \frac{1}{\pi^{(s)}_k} \tag{2.73}$$

with the sample inclusion probability $\pi^{(s)}_k = \pi^{(s')}_k \cdot \pi^{(s|s')}_k$ of element k in sample s. This probability is the product of the inclusion probability in sample s' and the conditional probability of inclusion in s given element k is in s'. It is this conditional probability $\pi^{(s|s')}_k$ where auxiliary information on x from s' can be incorporated, for instance, to stratify s' into H strata with respect to variable x before s is drawn. For example, in the first phase, an SI-sample s' of size n' is drawn from U to observe x. The number n'_h of units of s' that belong to stratum h is random ($h = 1, \ldots, H$). In the second phase, an STSI-sample $s = \{s_1, \ldots, s_H\}$ with stratum sample sizes $n_1, \ldots, n_H$ is drawn from $s' = \{s'_1, \ldots, s'_H\}$. In this case of two-phase sampling (SI/STSI), $\pi^{(s')}_k = \frac{n'}{N}$ applies for $k \in U$, and $\pi^{(s|s')}_k = \frac{n_h}{n'_h}$ applies for $k \in s'_h$. Consequently, the HT-type estimator (2.73) is given by

$$t_{\text{HT,SI/STSI}} = \frac{N}{n'} \cdot \sum_{h=1}^{H} n'_h \cdot \bar{y}_{s_h}$$

Fig. 2.15 Generating a
pseudo-population for the
HT-type estimator of a total in
two-phase sampling

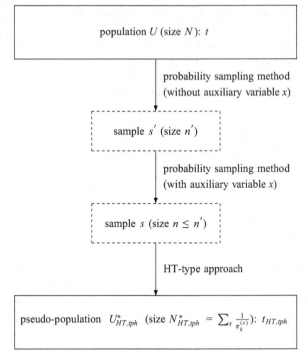

with $\bar{y}_{s_h}$, the mean value of y in s_h. Hence, the relative stratum size $\frac{N_h}{N}$ of stratum h in U is estimated by the relative stratum size $\frac{n_h'}{n'}$ of this stratum in the first phase sample s' ($h = 1, \ldots, H$).

The unbiased estimator $t_{\text{HT,tph}}$ of parameter t is intuitively conceivable as the generation of a pseudo-population $U^*_{\text{HT,tph}}$ for U with respect to the estimation of t (see Fig. 2.15). For this purpose, each sample unit of s is replicated $\frac{1}{\pi_k}$ times. Hence, $U^*_{\text{HT,tph}}$ is of size $N^*_{\text{HT,tph}} = \sum_s \frac{1}{\pi_k^{(s)}}$ and the estimator $t_{\text{HT,tph}}$ sums up all of the y^*-values constituted by these replicated values in $U^*_{\text{HT,tph}}$:

$$t_{\text{HT,tph}} = \sum_{U^*_{\text{HT,tph}}} y_k^*.$$

By using the auxiliary variable x from the first sample, pseudo-population $U^*_{\text{HT,tph}}$ may give better results with respect to the estimation of t, when compared with a pseudo-population generated without applying x at the same sample size n.

Another way to incorporate the auxiliary information observed in the phase one sample s' is to include it in the estimation phase of the second sample. This can be done, for instance, by a ratio-type estimator $t_{\text{rat,tph}}$. For this purpose, the total $t^{(x)}$ of auxiliary variable x is estimated by

$$t^{(x)}_{\text{HT},s'} = \sum_{s'} x_k \cdot \frac{1}{\pi^{(s')}_k},$$

its HT estimator in s'. The estimation of the total t of study variable y in the subsequent second sample s can be done by

$$t_{\text{rat,tph}} = t_{\text{HT,tph}} \cdot \frac{t^{(x)}_{\text{HT},s'}}{t^{(x)}_{\text{HT,tph}}}, \tag{2.74}$$

where $t_{\text{HT,tph}}$ and $t^{(x)}_{\text{HT,tph}}$ are the HT-type estimators of t and $t^{(x)}$ in s. Compared to the estimator t_{rat} according to (2.30), in (2.74), the total $t^{(x)}$ is substituted by its HT estimator calculated in the phase one sample s'. This means that in contrast to the ratio estimator t_{rat}, the HT-type estimator $t_{\text{HT,tph}}$ of t is corrected by the ratio of two different estimators of $t^{(x)}$. This corresponds to the generation of a pseudo-population $U^*_{\text{rat,tph}}$ with elements created by a $\left(\frac{1}{\pi^{(s)}_k} \cdot \frac{t^{(x)}_{\text{HT},s'}}{t^{(x)}_{\text{HT,tph}}} \right)$ times replication of each y_k-value of s. This results in a size of $N^*_{\text{rat,tph}} = \sum_s \frac{1}{\pi^{(s)}_k} \cdot \frac{t^{(x)}_{\text{HT},s'}}{t^{(x)}_{\text{HT,tph}}}$. The estimator $t_{\text{rat,tph}}$ sums up all of the y^*-values constituted by these clones in $U^*_{\text{rat,tph}}$:

$$t_{\text{rat,tph}} = \sum_{U^*_{\text{rat,tph}}} y^*_k.$$

Again, using information on x from s', the two-phase sampling design may be more efficient than a one-phase design not incorporating this information.

The variances of both examples of estimators from the two-phase samples consist of two summands. The first one is the variance that would occur if y is obtained in the first sample s'. The second one corresponds to the additional inefficiency from subsampling in the phase two sample s (cf., for instance, Särndal et al. 1992, p. 348).

Chapter 3
Nonresponse and Untruthful Answering

3.1 Introduction

Classical sampling theory considers only sampling errors and the effects of different sampling designs on this type of errors. Therefore, it can be said to be a pure full response theory with no place for nonresponse or untruthful answers. However, the practice of surveys does not comply with these assumptions. Nonresponse and untruthful answering are sources of so-called non-sampling errors. This term implies that such errors can also occur in a census.

With respect to the variable under study, in the presence of nonresponse, caused by unavailability of survey units, their refusal to participate at all, or their noncooperation on certain items, the set s of n sampling units drawn from U applying a probability sampling method without replacement is decomposed into a response set s_r ($s_r \subseteq s$) of size n_{s_r} and a nonresponse set s_m of n_{s_m} missing values ($s = s_r \cup s_m$, $s_r \cap s_m = \emptyset$, $n = n_{s_r} + n_{s_m}$). Additionally, especially for sensitive subjects, such as harassment at work, domestic violence, or drug use, the response set s_r is further divided into a set s_t of truthful answers ($s_t \subseteq s_r$) of size n_{s_t} and a set s_u of untruthful answers of size n_{s_u} ($s_r = s_t \cup s_u$, $s_t \cap s_u = \emptyset$, $n_{s_r} = n_{s_t} + n_{s_u}$) (see Fig. 3.1).

Hence, the HT estimator t_{HT} (2.4) for the total t of variable y is decomposed into

$$t_{\mathrm{HT}} = \sum_{s_t} y_k \cdot \frac{1}{\pi_k} + \sum_{s_u} y_k \cdot \frac{1}{\pi_k} + \sum_{s_m} y_k \cdot \frac{1}{\pi_k}. \tag{3.1}$$

For $s_u = \emptyset$ and $s_m = \emptyset$ only, Eq. (3.1) reduces to the full response HT estimator (2.4). The total absence of nonresponse and untruthful answering becomes a special case of (3.1). When set s_u is nonempty, for the summand over s_u in the middle of (3.1), "fakes" of the true values y_k are unknowingly observed instead of the true values ($k \in s_u$). Moreover, for the last summand, no observations y_k are available at all ($k \in s_m$).

© Springer International Publishing Switzerland 2015
A. Quatember, *Pseudo-Populations*, DOI 10.1007/978-3-319-11785-0_3

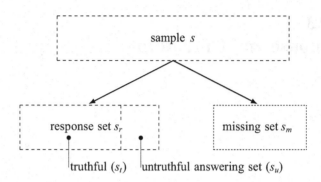

Fig. 3.1 The decomposition of a sample in the presence of nonresponse and untruthful answering

To analyze only the response set s_r as if it were the sample s and there were no untruthful answers in it is referred to as "available cases analysis." This leads to a biased estimator with an unknown extent of bias if the respondents differ from the nonrespondents with respect to the variable under study and $s_u \neq \emptyset$ applies. Consequently, this will massively affect the validity of confidence intervals. One must not ignore the fact that the set s_r is only a sample of those members of population U who are willing to participate and deliver the information asked in the survey either truthfully- or untruthfully. An example of the bad effect of available cases analysis is probably the persistent underestimation of the voting proportions of political parties belonging to the far left or right wings of the political spectrum in opinion polls in Europe.

Naturally, the best way to circumvent these problems is to avoid both nonresponse and untruthful answering. Even the most sophisticated method of compensation for nonresponse cannot be as good as the observation of the true value. Empirical social researchers have been considering survey design features that affect the quality and quantity of responses. Beatty and Herrmann (1995), for instance, referred to the respondent's decisions to respond and to provide a correct answer as "formulating communicative intent" (p. 1007). Respondents with negative communicative intent may believe, for instance, that their correct answer is socially undesirable (cf. Beatty and Herrmann 2002, p. 75). Groves et al. (2004) listed several strategies that have the potential to overcome such a negative intent (cf. p. 189ff). Some of these aspects may only have an effect on the nonresponse rate, whereas others may also influence the willingness to deliver the true values of the variables asked.

Among the aspects that mainly affect the quantity of replies is the survey sponsor. There is strong evidence that higher cooperation rates can be attained in surveys conducted by a public sector institution, such as a university, than in surveys by a market or opinion research institute (cf. Heberlein and Baumgartner 1978, p. 450ff). Also, the duration of the data collection phase of a survey has an influence on the response rates because of its relation to the effort planned to access the sampling units (cf. James and Bolstein 1990, p. 350ff). According to Botman and Thornberry

(1992), the length of the experiment or questionnaire is also among the survey design features with an impact on the respondents' willingness to participate (cf. p. 310).

The survey design features that clearly affect both the quantity and the quality of information asked from the respondents are strongly related to the sampling units' concerns about "data confidentiality" and their "perceived protection of privacy." The first term refers to the respondents' desire to keep their replies out of the hands of uninvolved persons and the second to their wish to withhold information from absolutely anybody. Singer et al. (1993, 2003), reporting on two successive U.S. population surveys, observed that the higher these concerns are, the lower the probability of the respondent's participation in the survey (cf. p. 470ff and p. 375ff). However, a stronger assurance of confidentiality does not seem to produce higher response probabilities (cf. Dillman et al. 1996, or the meta-analysis in Singer et al. 1995). In fact, it turns out that confidentiality assurances have a positive effect on response quantity and quality only when sensitive topics are asked. On the contrary, for non-sensitive questions such an assurance could be counter-productive (cf. Singer et al. 1995, p. 71ff). The method of data collection affects the nonresponse rate and the untruthful answering rate, especially for sensitive subjects (cf. the review of different methodological studies in Tourangeau and Smith 1996, p. 277ff). For face-to-face and telephone surveys, Kreuter (2008) lists several strategies to reduce the interviewers' effects on quality and quantity of the responses such as training and supervision (cf. Kreuter 2008, p. 371). Advance and persuasion letters also play a role in this context. The respondent's motivation to cooperate may increase if these letters emphasize the importance of the study and cater to the sample unit's needs (cf. Traugott et al. 1987, p. 529ff).

Another possibility to overcome a negative communicative intent is the offering of incentives. Monetary incentives appear to have a greater effect on willingness to participate in a survey than nonmonetary gifts. Further, prepayment incentives have a greater effect than promised incentives (cf. the results from experiments or from different meta-analyses of experiments that implemented incentives to increase the survey units' cooperation, in Church 1993; James and Bolstein 1990; Singer et al. 1999, 2000). In the context of the overall field costs of a survey, Singer (2002) reported on the cost-effectiveness of prepaid monetary incentives (cf. p. 174). There is also some evidence that incentives do not increase response rates at the expense of response quality (cf. Singer et al. 2000, p. 178ff).

Dillman (1978) was the first to integrate all these survey design features and all other survey-related details into one system, the "Total Design Method," to improve the capability to obtain acceptable response rates. To overcome the "one size fits all" approach of this method, Dillman (2000) proposed the "Tailored Design Method," in which he modified his principles by adapting them to various survey aspects to satisfy the needs of both the respondents and the survey.

A system that allows to react on paradata concerning different design features during the survey is termed "responsive survey design" (cf. Groves and Heeringa 2006). Such information might indicate the effect of sample clustering, offered incentives, or repeated call-backs on the accuracy of the estimates and also costs.

An example of such a design is the purely statistical responsive design of two-phase sampling (see Sect. 2.8.4).

When a sufficient degree of these features is applied in the survey's design phase, it might be reasonable to assume that all the answers given are at least truthful ($s_r = s_t$). Under this model, Eq. (3.1) reduces to

$$t_{\mathrm{HT}} = \sum\nolimits_{s_r} y_k \cdot \frac{1}{\pi_k} + \sum\nolimits_{s_m} y_k \cdot \frac{1}{\pi_k}. \qquad (3.2)$$

Looking at Eq. (3.2), it is obvious that in the presence of nonresponse, there exist two approaches to the estimation of parameter t by t_{HT}: Either one tries to estimate t only on the basis of the observations in the response set s_r, or the second sum of (3.2) over the missing set s_m has to be estimated (see Sects. 3.2 and 3.3).

For both approaches to be efficient, it is necessary to model the underlying nonresponse mechanism. The selection of a probability sample s drawn from U follows a given sampling scheme. The selection of the response set s_r from s follows an unknown nonresponse mechanism that determines the probabilities of certain sets s_r (or, vice versa, missing sets s_m) for a given sample s. Little and Rubin (2002) distinguished between three types of nonresponse mechanisms (cf. p. 11ff). Data are denoted as "missing completely at random" (MCAR) if the willingness to participate depends neither on y itself nor on an observable auxiliary variable vector $\mathbf{x}$. The available cases analysis, which simply ignores nonresponse, assumes MCAR.

If the survey units' willingness to participate in the survey depends on the observable auxiliary information $\mathbf{x}$ but not on the variable y under study, the nonresponse mechanism is called "missing at random" (MAR). If this model actually explains the nonresponse mechanism, such behavior of respondents can be compensated because the necessary auxiliary information for the modeling of the nonresponse mechanism is available for all sampling units. Therefore, such a situation (as well as the MAR-case) is called "ignorable" nonresponse (cf., for instance, Lohr 2010, p. 339). Various studies comparing census results from respondents and nonrespondents in surveys of official statistics have documented higher nonresponse rates for individuals belonging to the following groups: city dwellers, singles, couples without children, young and old persons, people who are divorced or widowed, people with a low level of education, and self-employed persons (cf., for instance, Holt and Elliot 1991, p. 334, or Bethlehem 2002, p. 285).

Eventually, if the response probability of a survey unit depends on the missing data, the nonresponse mechanism is called "not missing at random" (NMAR). In this case, a model cannot completely account for the nonresponse. In practice, in most surveys nonresponse is likely to be of this last kind. Nonetheless, auxiliary information can help to at least reduce the bias introduced in the estimation process by the "missings."

3.2 Weighting Adjustment

To be able to estimate the total t of variable y with (3.2) solely on the basis of the given responses y_k ($k \in s_r$), the design weights $\frac{1}{\pi_k}$ of the elements belonging to the response set s_r have to be adjusted. This is done with the help of the response probability ω_k of unit k of set s_r, leading to the adjusted weights $\frac{1}{\pi_k} \cdot \frac{1}{\omega_k}$. For $\omega_k = 1 \; \forall \; k \in U$, the adjusted weight equals the design weights, and we have the classical sampling theory approach assuming full response. However, for $s_m \neq \emptyset$, the individual sample elements have to bear a heavier representation burden because of the nonrespondents.

Alternatively, this can be described in the following way applying the concept that was introduced in Sect. 2.3: Because of nonresponse, a pseudo-population U_{HT}^* formed in the usual way by replicating each y_k in the response set only $\frac{1}{\pi_k}$ times is expected to be too small. Hence, the idea is to generate a higher number of clones from the observations in s_r and increase the representation burden of each responding unit. This weighting adjustment (W) results in

$$t_W = \sum_{s_r} y_k \cdot \frac{1}{\pi_k} \cdot \frac{1}{\hat{\omega}_k}. \tag{3.3}$$

Estimator t_W estimates parameter t unbiasedly if the ω_k-values are estimated correctly by the $\hat{\omega}_k$-values. Of course, this is the crucial assumption with this estimator. To determine these estimated probabilities $\hat{\omega}_k$, the underlying nonresponse mechanism has to be modeled as discussed at the end of the previous section. The question that needs to be addressed is: How does the model fit the reality? The estimator t_W is model-based because the effect of a wrong nonresponse model is a biased estimator with an unknown extent of bias. With the pseudo-population U_W^* built up by $\left(\frac{1}{\pi_k} \cdot \frac{1}{\hat{\omega}_k} \right)$ clones of each response set unit k, t_W can be represented by

$$t_W = \sum_{U_W^*} y_k^*,$$

the total of all replicated values y_k^* in U_W^* (see Fig. 3.2). This shows the importance of a good model for the response mechanism because the estimated probabilities $\hat{\omega}_k$ together with the sample's design weights determine the composition of U_W^*.

For a MCAR mechanism, the response probabilities are equal for all elements of s: $\omega_k = \omega \; \forall \; k \in s$. In estimating this constant probability considering this assumption about the nonresponse mechanism, the response rate $\frac{n_{s_r}}{n}$ will make sense. An assumed MAR mechanism regards the selection of the response set s_r from the sample s as a stratified simple random selection from s according to a known one- or more-dimensional stratification variable. The response probabilities are considered to be equal for all elements within the same "response stratum" s_{r_h} of H such strata in the sample: $\omega_k = \omega_{s_h} \; \forall \; k \in s_h$ ($h = 1, \ldots, H$). Hence, the use of the response rate $\frac{n_{s_r,h}}{n_h}$ in stratum h as an estimator of the response probability ω_h in

Fig. 3.2 Generating a
pseudo-population for
weighting adjustment

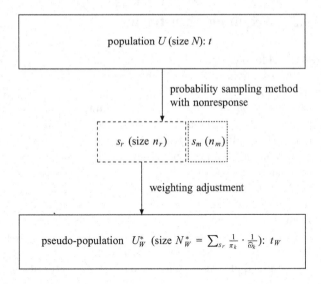

stratum h will be a logical consequence of the nonresponse model ($n_{s_{r,h}}:=$ number
of respondents in the sample of stratum h).

The variance expressions developed for two-phase sampling can be used in this
context (cf., for instance, Särndal et al. 1992, p. 581), viewing the sample selection
as the first phase and the sampling units' response behavior as the second phase as
it was presented in Sect. 2.8.4.

3.3 Data Imputation

The other possibility to compensate for nonresponse is to estimate the second sum of
(3.2). This is done by an imputation of a substitute value y_k^i for each unobserved y_k of
study variable y ($k \in s_m$). With respect to y, this results in a sample substituting y by
an "imputed variable" y^I consisting of actually observed and imputed values. This
yields a sample s_I with the same n elements as s, but with a different assignment:
$s_I \rightarrow \{y_1^I, \ldots, y_n^I\}$. Hence, the total estimator t_I based on the HT principle is given by

$$t_I = \sum_s y_k^I \cdot \frac{1}{\pi_k} \tag{3.4}$$

with

$$y_k^I = \begin{cases} y_k & \text{if } k \in s_r, \\ y_k^i & \text{otherwise.} \end{cases}$$

Fig. 3.3 Generating a
pseudo-population for data
imputation

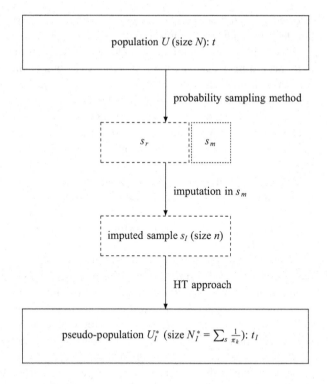

The idea behind this estimator of t is to generate a pseudo-population U_I^* from the
"imputed sample" s_I, of which either the observed value y_k or the imputed value y_k^i
of each sample unit $k \in s$ is replicated $\frac{1}{\pi_k}$ times ($k \in s$). For this purpose, information
on the auxiliary variables $\mathbf{x}$ available for each survey unit in the missing set s_m is
used for a reasonable imputation of variable y instead of excluding the whole unit
from the analysis (Fig. 3.3). The quality of U_I^* with respect to the estimation of t
depends on the quality of the estimation of $\sum_{s_m} y_k \cdot \frac{1}{\pi_k}$ in (3.2) by $\sum_{s_m} y_k^i \cdot \frac{1}{\pi_k}$, which
is a part of the sum (3.4). For measures concerned with the statistical relation of
variables such as the correlation between two quantitative variables, for instance,
the quality of U_I^* with respect to the estimation of these parameters depends, of
course, on the quality of the estimation of each missing value y_k by its imputed
estimate y_k^i ($k \in s_m$).

Different imputation methods are used to determine the substitute values y_k^i (see,
for instance, Little and Rubin 2002, Chaps. 4 and 5, or Lohr 2010, Sect. 8.6).
Deductive imputation, for instance, uses logical relationships between variables
to impute the correct variable value y_k by deducing it from the values known for
other variables x or from other sources. The term "cold-deck imputation" describes
a method in which the values y_k^i are imputed from other sources, such as a previous
survey.

Other techniques are based on different assumptions about the nonresponse
mechanism. "Hot-deck" procedures substitute missing values y_k in s_m for y_k^i-values

selected from donors in s_r of the same survey. One such procedure is "random hot-deck imputation." In this type of data imputation, the donor delivering the variable value to the recipient is randomly chosen from the same group of survey units according to an auxiliary variable **x**, to which the recipient belongs with respect to his or her nonresponse probability when the MAR model is assumed ("random hot-deck imputation within classes"). When a MCAR model is assumed, the donor is selected randomly from the entire response set ("random overall hot-deck imputation"). When the correlation structure of variables is also of interest, all missing values of a certain survey unit are replaced by the observed values of the same donor. For "nearest neighbor imputation," the nearest survey unit with respect to some defined measure of distance provides his or her value y_k to the unit with the missing value. In "sequential hot-deck imputation," the idea is that in the case of survey units ordered in a list according, for instance, to a regional variable, the unit observed right before the unit with the missing value will likely have a similar variable value.

However, the most natural imputation procedure is, of course, "regression imputation." In the deterministic case, the missing values of a variable y are replaced by the regressand of a regression equation calculated from data of the response set s_r using as independent regressors the auxiliary variables also available for the units in s_m. In the stochastic version of regression imputation, a stochastic error term is added to the regressand. The respondents' "mean imputation" can be seen as a special case of regression imputation. It can also be applied in a stochastic version, in which the imputations y_k^i are generated from an assumed probability distribution with the mean and variance calculated from the responding units. The technique of regression imputation can also be applied overall or within classes with the assumption of a MCAR and a MAR nonresponse mechanism, respectively.

Applying an imputation method, for which no numerical solution exists with regard to the estimation of the variance $V(t_I)$ of the estimator t_I, this variance can be estimated, for example, by the bootstrap method as discussed in Sect. 5, taking into account the imputation technique applied. Shao and Sitter (1996) presented this integration of data imputation into the bootstrap scheme. In the presence of values imputed into the sample in place of the missing values, treating the sample as if there were only true values and applying standard variance estimators will surely underestimate the real variance of the estimator. To incorporate imputation into the bootstrap technique (see Fig. 5.2), the originally used imputation method, whether deterministic or stochastic, is also applied in each of the B bootstrap resamples drawn from the generated bootstrap population. For this purpose, the imputed values must be identifiable in the data set. Then, the bootstrap estimators $\hat{\theta}_1^*, \ldots, \hat{\theta}_B^*$ are calculated in the B bootstrap samples considering the re-imputed values, and the variance can be estimated by (5.1). In this way, the additional inaccuracy due to data imputation of missing values is incorporated into the variance estimation process.

A different method that accounts for the imputation inaccuracy when a standard variance estimator is available and also applies the pseudo-population concept is "multiple imputation" (cf. Rubin 1987). It uses D imputations y_k^i for each missing

Fig. 3.4 Generating a
pseudo-population for
multiple imputation

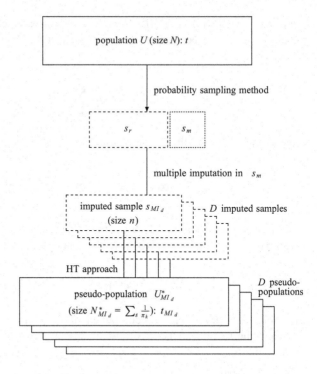

y_k in s, resulting in D differently imputed data sets s_{MI_d} ($d = 1, \ldots, D$). On the one
hand, the mean value of the calculated estimator of interest in each of these D data
sets can serve as an estimator for the parameter. On the other hand, the average of
the D standard variance estimates provides an estimate of the full response variance
of the estimator, whereas the additional inaccuracy caused by the imputation process
can be estimated by the variance between these data sets.

For the estimation of t, this means in the present context that from the D
imputed data sets s_{MI_d}, different pseudo-populations $U^*_{MI_1}, \ldots, U^*_{MI_D}$ are generated
by replicating the observed or the imputed value of sample unit k of imputed sample
$s_{MI_d} \frac{1}{\pi_k}$ times (see Fig. 3.4). This results in the replication variable y^{I*} in each $U^*_{MI_d}$.
In each of the D pseudo-populations, the estimator

$$t_{MI_d} = \sum_{U^*_{MI_d}} y^{I*}_k$$

is calculated according to (3.4) resulting in a set $t_{MI_1}, \ldots, t_{MI_D}$ of estimators with a
mean value

$$\bar{t}_{MI} = \frac{1}{D} \cdot \sum_{d=1}^{D} t_{MI_d}. \tag{3.5}$$

The average $\bar{t}_{MI}$ serves as the multiple imputation (MI) estimator for t.

Further, let $\hat{V}(t_{\mathrm{HT}_1}), \ldots, \hat{V}(t_{\mathrm{HT}_D})$ be D standard variance estimates treating the values y_k^i of the imputed sample $s_{\mathrm{MI}_1}, \ldots, s_{\mathrm{MI}_D}$ as if they were the true y_k-values. We define

$$\overline{V}(t_{\mathrm{HT}}) = \frac{1}{D} \cdot \sum_{d=1}^{D} \hat{V}(t_{\mathrm{HT}_d})$$

as an estimator of the full response variance of t_{HT}. Then,

$$\hat{V}(t_{\mathrm{MI}}) = \frac{1}{D-1} \cdot \sum_{d=1}^{D} (t_{\mathrm{MI}_d} - \bar{t}_{\mathrm{MI}})^2$$

is an estimator of the variance between the D data sets caused by imputation. Hence, an estimator of the total variance of $\bar{t}_{\mathrm{MI}}$ is given by

$$\hat{V}(\bar{t}_{\mathrm{MI}}) = \overline{V}(t_{\mathrm{HT}}) + \frac{D+1}{D} \cdot \hat{V}(t_{\mathrm{MI}}) \tag{3.6}$$

(cf. Rubin 1987, p. 76). The first component of (3.6) is the estimator of the sampling error, and the second of the additional inaccuracy caused by data imputation for nonresponse.

3.4 Combining Data Imputation and Weighting Adjustment

In fact, weighting adjustment and data imputation are two supplementing methods to compensate for nonresponse. Usually, weighting adjustment is applied to compensate for unit nonresponse, whereas data imputation is used in the presence of item nonresponse. In practice, both types of nonresponse are natural in surveys. Therefore, the two techniques are often applied in succession, starting with the compensation of item nonresponse by data imputation followed by that of unit nonresponse by weighting adjustment. This means that an imputed value y_k^i is calculated for each missing value y_k in the item nonresponse subset $s_{m,\mathrm{item}}$ of s_m only ($s_{m,\mathrm{item}} \subseteq s_m$). For $s_{m,\mathrm{item}} \subset s_m$, to compensate also for the unit nonresponse that has occurred in the set $s_{m,\mathrm{unit}} = s_m - s_{m,\mathrm{item}}$, the imputation is followed by an adjustment of the design weights $\frac{1}{\pi_k}$ of elements k belonging to the imputed sample $s_I = \{s_r, s_{m,\mathrm{item}}\}$ (see Fig. 3.5). When data imputation and weighting adjustment (IW) are conducted in succession, this leads to the following estimator of t:

$$t_{\mathrm{IW}} = \sum_{s_I} y_k^i \cdot \frac{1}{\pi_k} \cdot \frac{1}{\hat{\omega}_k}. \tag{3.7}$$

Fig. 3.5 Generating a pseudo-population for the two-step process of data imputation and weighting adjustment

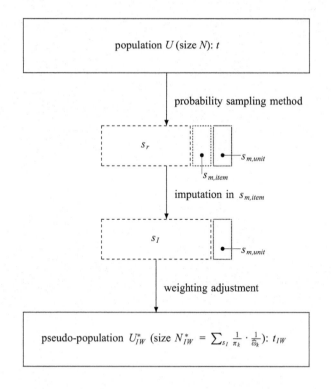

From the point of view of the pseudo-population concept, this means that the y_k^I-values of the imputed sample s_I are replicated $\frac{1}{\pi_k} \cdot \frac{1}{\bar{\omega}_k}$ times according to the rules of weighting adjustment to generate a pseudo-population U_{IW}^* of size $\sum_{s_I} \frac{1}{\pi_k} \cdot \frac{1}{\bar{\omega}_k}$. Hence, with the replications y^{I*} in U_{IW}^*, the estimator t_{IW} can be represented by

$$t_{\text{IW}} = \sum_{U_{\text{IW}}^*} y_k^{I*}.$$

Chapter 4
Simulation Studies in Survey Sampling

4.1 The Basic Simulation Approach

An example of an application of the concept of pseudo-populations in research comes from the field of computer simulation studies. This term describes a process of conducting experiments, which are actually part of real life, on the computer. In the context of statistics, simulation studies are applied when mathematical derivations of the statistical properties of a method are cumbersome or not available at all.

In finite population sampling, such studies are conducted, for example, when the given complex sampling design does not allow for the formal calculation of the bias or variance of an estimator $\hat{\theta}$ for an interesting parameter θ. Other possible issues are, for instance, the estimation of the actual coverage rate of an approximate confidence interval under normal assumption or the robustness evaluation of a particular estimator under assumption violations. Possible aims may be the strengthening of the confidence in the quality of a proposed method or a comparison of the efficiency of different methods.

Naturally, the perfect starting point for these purposes would be the knowledge of the original population U. Then, the used sampling method can be applied to these data again and again to draw as many samples as possible on the computer. In each of these B simulation runs the estimator under investigation is calculated. For large B, the empirical distribution of these B estimates approximates the true sampling distribution over all samples shown in Fig. 2.1 (see, for example, the simulation study described in Sect. 5.3).

Unfortunately, population data are rarely available because they often simply do not exist or they are not open to the public for reasons of data protection (see Sect. 7). In such cases, a plausible (or synthetic) population U_{sim}^* may be generated for the simulation purposes from different sources such as the data of a sample survey and other available information about the population (see Fig. 4.1). Such a population shall serve as a close-to-reality pseudo-population for the non-observable original

© Springer International Publishing Switzerland 2015

A. Quatember, *Pseudo-Populations*, DOI 10.1007/978-3-319-11785-0_4

Fig. 4.1 Generating a plausible population for a simulation study in the field of sample surveys

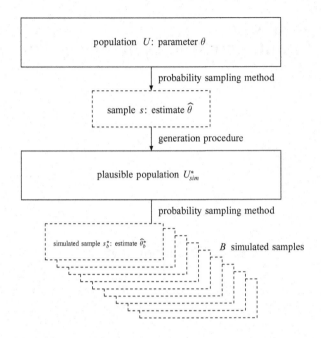

population U so that the simulation results in the samples from U^*_{sim} are comparable to those which would occur in samples from U.

For the pseudo-population to be plausible, different generation procedures may be considered depending on the available data and the specific requirements of the simulation study (cf. Münnich and Schürle 2003, or Alfons et al. 2011a). Typically, an adequate pseudo-population should reflect as close as possible the structure of the original population with respect to stratification and clustering in the sample. The number of clusters within each stratum can be generated either on the basis of a given distribution or known information about the cluster population itself. To avoid impossible combinations of important variables within the clusters, realistic combinations may be drawn from survey data. The other variables needed for the simulation can be generated from conditional distributions, which are calculated from a given modeled or true multivariate distribution (cf. Templ et al. 2011, p. 12ff). In addition, the original nonresponse mechanism and other survey characteristics must be modeled if they are also part of the investigation (cf. Alfons et al. 2011b, p. 7f).

4.2   An Example of a Simulation Study Based on a Generated Plausible Population

An example of such a simulation study is a secondary analysis of the Austrian data from PISA survey 2009. The Programme for International Student Assessment (PISA) is currently one of the most important and most influential statistical surveys in the field of educational sciences. It is conducted on a triennial basis by the Organization for Economic Co-operation and Development (OECD). Its topic is the measurement of the abilities of the current birth cohort of 15- to 16- year-old students with respect to their competencies in reading, mathematics, and natural sciences (cf. here and in the following, OECD 2012). The results allow for cross-national comparisons of these skills as effects of the different educational systems and for the measurement of the effects of changes within national systems across time.

The PISA survey 2009, for example, was conducted in the 34 OECD member states and further in 31 countries outside of the OECD. The target population this year was the group of about 26 million students born in 1993, attending school in 65 countries. In almost all countries, the PISA survey is not a full, but a sample survey of the target population. Furthermore, the applied item-response model, which calculates the distribution of competence of a unit and not a single value, is an instrument deliberately chosen to document the inaccuracy of the measurement process. For these two reasons, the PISA-results, which are the mean values of the measured skills within each country, clearly, are only estimates of the related unknown population parameters.

The calculation of the accuracy of these estimators of the population means is nontrivial because the sampling method is of high complexity. To demonstrate this complexity, in the following, the main features of the setup of the national PISA survey 2009 in Austria serves as the representative for all participating countries: The first-stage selection units contained in the sampling frame are schools with pupils of the target population. These schools were explicitly stratified according to the two stratification variables "region" and "school type," giving a total of 32 strata. The sample number of schools per stratum corresponds to the relative stratum size with respect to the target population. The selection probabilities of different schools depend on the size of the schools defined as the number of students belonging to the target population. Schools with more than 35 students of the target population are "large schools." The selection probabilities of such schools are proportional to their school sizes. "Small schools" of size 35 or less have sample selection probabilities proportional to a school size of 35.

To generate a school sample with such first-order selection probabilities on school level, the systematic sampling mode as described in Sect. 2.4.1 is applied. However, the ordering of the schools is not done randomly, but by specific variables. In Austria, these are "province," "percentage category of girls attending the school," and "school size." A systematic random selection from such a list introduces an "implicit stratification" with respect to these variables into the sampling procedure.

Within schools drawn by this selection process, the selection of the actual test persons differ depending on the school size. In large schools, 35 test students are selected randomly according to a systematic sampling mode with uniform selection probabilities from a list of sampling units ordered by "level of education," "sex," and "age." In small schools, all students are tested.

This results in uniform first-order selection probabilities for all students within the same stratum of the target population because, on the one hand, for all students

k in large schools i, $\pi_k = \kappa_i \cdot \pi_{k|i} = \frac{N_i m}{N} \cdot \frac{35}{N_i} = \frac{35m}{N}$ applies (see Sect. 2.4.6). On the other, for elements k attending a small school i, with conditional probability $\pi_{k|i} = 1$, these inclusion probabilities are also given by $\pi_k = \kappa_i = \frac{35m}{N}$. The resulting design weights of the N population units of the same stratum are uniformly given by $\frac{N}{35m}$. Hence, the entire PISA sample is approximately self-weighting.

Moreover, the test results of a test person in the three key competencies of the study are not single values describing his or her skills but posterior distributions of these three skills. These posterior distributions are calculated from prior distributions assigned to each student with respect to some socio-economic variables, which are corrected in light of the individual test results. To be able to calculate the mean values though, from the resulting posterior distributions, five "plausible values" are drawn from each one randomly under i.i.d. conditions (cf. Mislevy 1991).

The sampling method applied in the PISA survey can be described as a stratified two-stage random sampling with proportional allocation of primary sampling units to strata. Within each stratum, the selection of m schools at the first stage of the TS selection process is done randomly with implicit stratification by systematic selection and first-order sample inclusion probabilities proportional to school size. At the second stage of the TS scheme of a stratum, the secondary sampling units are chosen either randomly with implicit stratification by systematic selection and uniform sample inclusion probabilities or with probability one depending on the school size. The πPS sampling, implicit stratification, and systematic selection mode, together with the measurement procedure, make the calculation of an analytical variance estimator for the sample results more than cumbersome.

To estimate the sampling variance of the mean values per country and competence, in the original PISA survey, the resampling method of balanced repeated half sampling according to Fay (1989) and Judkins (1990) was adapted to the circumstances of the PISA sampling procedure (OECD 2012, p. 126ff). The additional inaccuracy due to the use of plausible values was implemented into the variance estimation as a measurement error.

In the secondary analysis of the Austrian PISA data of the 2009 survey, Quatember and Bauer (2012) investigated different tasks concerning the sampling process of the PISA survey on the bases of simulations. These issues were a comparison of the efficiency of different sampling methods, the design effect of the actual applied sampling method, and the validity of the approximate confidence intervals calculated under the normal assumption.

For the simulation purposes, in absence of the true population U, different methods for the generation of a bootstrap population as described in Sect. 5 and an adjustment to known population parameters are integrated into the generation procedure of a plausible population of pupils (see Fig. 4.1). This pseudo-population reflects the given structure of the Austrian school population with respect to school-types and corresponds exactly to the original target population of the PISA survey with regard to the number of schools and the number of students of both sexes within these schools. Hence, it can serve as basis for a close-to-reality simulation study with respect to the issues mentioned above.

From this pseudo-population, $B = 10,000$ samples are drawn for each sampling method included in the study to compare the effect of different sampling schemes on the efficiency of the survey results. Implicit stratification is ignored at both stages of the sampling process. For each test student of a simulation run, five plausible values are drawn from an assumed normal posterior distribution estimated from the actual five plausible values for each of the three skills under investigation. Because of the "imputation" of these five values for the "missing" true value of a test student, the method of multiple imputation (see Sect. 3.3) is applied to calculate a single estimator of the pseudo-population mean for each skill in each of the B simulated samples combining the ideas of Shao and Sitter (1996) and Rubin (1987). The distributions of these B estimators are interpreted as an approximation to the sample distributions of the estimators under the given design.

The mean values of the three competencies (reading, mathematics, and science) calculated in the simulation study by drawing resamples from a pseudo-population as previously described show results that are very close to the original results. This underpins the plausibility of the pseudo-population generated in this way.

Summarizing the most important results of this application of the pseudo-population concept in a simulation study regarding the Austrian PISA survey 2009, the actual PISA sampling method seems to provide good results from the sampling theory point of view considering, in particular, the disproportionate higher cost and time efforts of the SI method of sampling as the reference sampling scheme. This also applies to other sampling methods, to which the PISA sampling mode is compared.

The overall design effect [see (2.22)] of the true PISA design was about 4 for all three skills (for the details concerning the results of this secondary analysis, see Quatember and Bauer 2012, p. 540ff, and Bauer 2011). Moreover, it is shown for a sufficiently large stratum in the pseudo-population that the systematic selection of schools and students within sample schools may result in multi-modal sampling distributions, for which approximate confidence intervals based on the normal assumption are not valid.

Chapter 5
The Bootstrap Method in Survey Sampling

5.1   The Finite Population Bootstrap Approach Based on Pseudo-Populations

When no explicit variance formula is available and the calculations for Taylor linearization (cf., for instance, Wolter 2007, p. 230ff) are too cumbersome, so-called computer-intensive methods that use computer power instead of heavy calculations can be applied alternatively. One such procedure is the random group method (cf., for instance, Wolter 2007, Chap. 2). In this case, the sample drawn is divided into different nonoverlapping subsamples, called "random groups," according to the original sampling design. After calculating the original estimator of the parameter under study in each of the groups, the variance of these estimators serves as the basis for extrapolation regarding the variance of the estimator in the original sample. The calculations are truly simple, but for obvious reasons are often inefficient for complex surveys because the construction of subgroups according to the original sampling design might be difficult.

To overcome the problem of poor efficiency of the random group method, the balanced repeated half sampling was developed where a stratified sampling design with only two units per stratum is used as a sampling method. The basic idea is to generate half samples by the selection of one element per stratum. If this is done "in a balanced way," the estimators for the parameter of interest, calculated in each of the half samples, can be used to estimate the variance of the parameter estimate in the original sample (cf., for instance, Wolter 2007, Chap. 3). The method has been extended to other sampling schemes by applying adapted techniques of balancing the half samples.

Another technique of estimating the theoretical variance of an estimator is the bootstrap method. This strategy falls under the resampling methods. While another strategy, the jackknife method, generates resamples from the original sample, which consist of all but one or a certain number of elements of the original sample drawn, the basic bootstrap procedure generates resamples of the same size as the original

© Springer International Publishing Switzerland 2015
A. Quatember, *Pseudo-Populations*, DOI 10.1007/978-3-319-11785-0_5

sample. The reason why we concentrate on the bootstrap is that a major approach to the application of this procedure to finite population surveys is based on the idea of generating pseudo-populations that runs like a red thread through this work.

Lahiri (2003) described the bootstrap as "probably the most flexible and efficient method of analyzing survey data since it can be used to solve a variety of challenging statistical problems (e.g., variance estimation, imputation, small-area estimation, etc.) for complex surveys involving both smooth and non-smooth statistics" (p. 199). This technique was originally developed by Efron (1979) for the calculation of the sample distribution of an estimator $\hat{\theta}$ for the parameter θ of a probability distribution ϕ. For this purpose, a sample drawn according to the i.i.d. principle is observed (cf., for instance, Casella and Berger 2002, p. 207). The observed empirical distribution of a random variable y can be interpreted as the ML estimator of the true probability distribution ϕ of y (cf. Chao and Lo 1994, p. 391ff). Drawing i.i.d. resamples of the same size as the original sample from the empirical distribution, the true sample distribution of $\hat{\theta}$ is approximated by the theoretical distribution of the estimator calculated in all possible resamples (for the mathematical details see Shao and Tu 1995). This bootstrap distribution in turn can be approximated by the Monte Carlo approximation. For this purpose, a number of, say, B resamples is drawn. Within each of the B bootstrap samples $s_1^*, \ldots, s_B^*$, the estimator $\hat{\theta}_b^*$ is calculated in the same way that the estimator $\hat{\theta}$ was calculated in the original i.i.d. sample s ($b = 1, \ldots, B$). For a large B, the distribution of $\hat{\theta}_b^*$ is interpreted as an estimation of the sample distribution of $\hat{\theta}$. Hence, the theoretical variance $V(\hat{\theta})$ is estimated by the Monte Carlo variance estimator given by

$$\hat{V}_{\text{boot}}(\hat{\theta}) = \frac{1}{B-1} \cdot \sum_{b=1}^{B} \left(\hat{\theta}_b^* - \overline{\theta}^* \right)^2 \tag{5.1}$$

with

$$\overline{\theta}^* = \frac{1}{B} \cdot \sum_{b=1}^{B} \hat{\theta}_b^*,$$

being the mean value of estimators $\hat{\theta}_b^*$ from the B bootstrap samples. For approximately normally distributed $\hat{\theta}_b^*$ (or $\hat{\theta}$) values, this variance estimator can be used for the calculation of an approximate confidence interval. For a large B, also for non-normally distributed bootstrap estimators, a confidence interval can be calculated by applying the percentile method (Efron 1981, p. 317ff). This method directly uses the $(\alpha/2)$- and $(1 - \alpha/2)$-quantile of the observed distribution of the estimators $\hat{\theta}_b^*$ as the lower and the upper bound of the confidence interval, respectively (cf. Efron 1981).

With increasing computer power, this technique has also become attractive for finite populations surveys. In this context, an extension of Efron's technique has to consider complex sampling designs consisting of complex estimators and complex techniques of sampling without replacement at various stages of the sample process. For this purpose, different approaches are available in the relevant literature (cf., for instance, Shao and Tu 1995, p. 247ff, or Wolter 2007, p. 200ff). One of them rescales the observations in the resamples drawn with replacement from the original without-replacement sample in a way that the bootstrap variance (5.1) approximates the actual variance for a given sampling design (cf. Rao and Wu 1988). Another approach is to use the with-replacement bootstrap technique and adjust its bootstrap variance estimator to the parameter by choosing an appropriate size for the resamples (cf. McCarthy and Snowden 1985). Sitter (1992a) presented the Mirror-Match Method, in which subsamples of the original sample are drawn repeatedly according to the original sampling plan with a subsample size chosen to appropriately match the original variance of the estimator. Antal and Tillé (2011) discuss another approach, in which different with- and without-replacement resampling procedures are combined in such a way that the bootstrap variance estimator, calculated from resamples of the same size as that of the original without-replacement probability sample, under this combination of resampling schemes equals the interesting variance.

However, for a direct extension of the i.i.d. bootstrap to finite population sampling, the population U of N elements takes over the role of the unknown probability distribution ϕ. The population elements are characterized, as always, by their values y_k of the variable y under study and $\mathbf{x}_k$ of possible auxiliary variables $\mathbf{x}$ ($k = 1, \ldots, N$). Gross (1980) was the first to adapt this method to the specific case of SI sampling, but only with restriction of integer design weights $\frac{N}{n} \in \mathbb{N}$ (cf. Gross 1980, p. 184). For this purpose, from the SI sample s, a set-valued estimator U^* of the true population U of size N is generated according to the HT pseudo-population approach to SI sampling with $N^* = N$ (see Sect. 2.4.2 and Fig. 2.2). Hence, for this restriction, pseudo- (or in this case, bootstrap) population U^* equals U_{SI}^* and consists of $\frac{N}{n}$ replications of each element of the sample s with respect to the variables observed. The generated bootstrap population is the finite population of size N with the maximum likelihood regarding the sample drawn.

In the next step of the without-replacement bootstrap as proposed by Gross (1980) (see Fig. 5.1), B bootstrap samples $s_1^*, \ldots, s_B^*$ of size n are drawn from the bootstrap population by applying the original sampling method. In other words, the resamples are no i.i.d. samples of size n from the original sample s. Instead, the resampling process from U^* follows a poly-hypergeometric distribution. Hence, each of the n sample values $y_1, \ldots, y_n$ has the same probability $\frac{1}{n}$ of being chosen as the first value in the resample of same size n. After the first draw, the value, already drawn at the first step, has a probability of $\frac{N-n}{n(N-1)}$ for being chosen as the second element of the resample. The other $n - 1$ values of y in s, not selected as the first resample element, have a probability of $\frac{N}{n(N-1)}$ and so on. Generally, the value y_k observed in s has a probability

$$\frac{N - n \cdot h_{k,j-1}}{n \cdot (N - j + 1)}$$

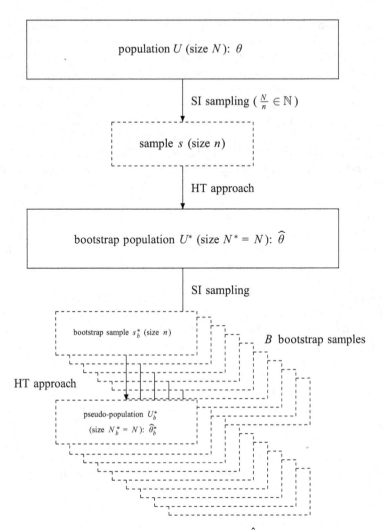

Fig. 5.1 Estimating the sampling distribution of an estimator $\widehat{\theta}$ applying the bootstrap method in SI sampling with integer design weights according to Gross (1980)

of being selected at the jth step of the selection of a resample from $U^* = U^*_{SI}$ ($j = 1, \ldots, n$). Therein, $h_{k,j-1}$ denotes the number of times the value y_k was already selected in the first $j - 1$ steps of the process to generate a resample ($h_{k,0} = 0$ $\forall\ k \in s$). This shows that such bootstrap populations need not be generated in reality. Obviously, the resampling process might as well be carried out by applying the probability mechanism described above directly to the sample s (for other procedures that can be applied with different sampling schemes instead of the physical generation of a bootstrap population, cf. Ranalli and Mecatti 2012).

These resamples form the basis for estimating the SI sampling distribution of the estimator $\hat{\theta}$ for the interesting parameter θ based on simulations. For this purpose, in each of the B resamples s_b^*, the estimator $\hat{\theta}_b^*$ has to be calculated ($b = 1, \ldots, B$). Considering, for instance, the estimation of parameter t of variable y by the HT estimator t_{SI}, this means that within each resample s_b^*, an estimate $t_{\text{SI}_b}^*$ is calculated. This comprises, in particular, the generation of B pseudo-populations $U_{\text{SI}_1}^*, \ldots, U_{\text{SI}_B}^*$, each of them estimating the bootstrap population U^* by replicating the resampling units of each of the bootstrap samples $s_1^*, \ldots, s_B^*$ $\frac{N}{n}$ times. Within these B pseudo-populations generated from the B resamples, the bootstrap estimators $t_{\text{SI}_1}^*, \ldots, t_{\text{SI}_B}^*$ are calculated, and their variance (5.1) serves as an estimator of the variance of t_{SI}. This variance estimator is approximately unbiased in large samples (cf., for instance, Chao and Lo 1985, p. 400, or Sitter 1992b, p. 139).

For general applicability in survey sampling, this idea had to be extended to

- non-integer design weights, and
- general probability sampling with arbitrary first-order inclusion probabilities.

In fact, the key to an efficient application of the bootstrap method in finite population sampling without replacement in this way is the generation of a pseudo-population U^* suitable as the basis for drawing the bootstrap samples with respect to the estimation problem to be solved. In the following section, an overview of various ideas in this respect is given with an application to the estimation of the total t of y. In Sect. 5.3, these suggestions are presented in a general framework.

5.2 An Overview of Different Finite Population Bootstrap Techniques

The most important step in this type of bootstrap approach to finite populations is the generation of an adequate bootstrap population U^*. Bickel and Freedman (1984), for example, addressed the question of the actual number of replications of each sample element when the constant design weight of the SI method is not an integer. They departed from the ML principle in the generation process used by Gross (1980) by generating two bootstrap populations, U_a^* and U_b^*, instead of one. The first consists of a number of i replicates of the n sampling units. Therein, i denotes the integer part of the ratio $\frac{N}{n}$. For the second, each sample value y_k is cloned $(i + 1)$ times. This creates pseudo-populations of size $N_a^* = n \cdot i$ and $N_b^* = n \cdot (i+1)$, respectively. Both pseudo-populations have the same estimated cumulative distribution function of y. Note that for $\frac{N}{n} \notin \mathbb{N}$, neither N_a^* nor N_b^* equals the true size N of U. Then, one of these populations is selected according to "an artificial randomization" (Rao and Wu 1988, p. 237) with probabilities appropriately calculated to compensate for incorrect population sizes with respect to the unbiased estimation of the variance of t_{SI}. After randomly choosing the actual bootstrap population, its design weights are adapted to $\frac{N_a^*}{n}$ or $\frac{N_b^*}{n}$, both being integers, and the resampling procedure is applied. It can be shown that the estimator for the variance of the HT estimator is approximately unbiased for large n (cf. Bickel and Freedman 1984, p. 474, or Chao and Lo 1985, p. 401).

After modifying the size of the resamples, the replication factors, and the probabilities of selecting one of the two bootstrap populations of Bickel and Freedman (1984) for SI sampling, Sitter (1992b) also applied the method to STSI, TSSI and, with replication factors corresponding to the specific design weights of πPS sampling (see Sect. 2.4.1), to πPS sampling according to Rao–Hartley–Cochran model (cf. Cochran 1977, p. 266f). Such an application of the bootstrap method to a sampling design dividing the given population into strata and clusters must reflect this given structure in the bootstrap population (cf. Chao and Lo 1994, p. 398).

Booth et al. (1994) presented another answer to the central question of how the finite bootstrap population should be generated from an SI sample with non-integer $\frac{N}{n}$-values: A number of i clones of the original SI sample s are combined with an SI subsample of size $r \cdot n = N - i \cdot n$ drawn from s $\left(\frac{N}{n} = i + r\right)$. Considering the randomness of the generation of the $r \cdot n$ supplementary elements, C pseudo-populations $U_1^*, \ldots, U_C^*$ of correct size N are created, which serve as the basis for the resampling process. In this process, B SI resamples of the original size n are drawn from each of these C pseudo-populations. The variance of an estimator $\hat{\theta}$ for a parameter θ such as the total t within these $C \cdot B$ bootstrap samples gives an approximately unbiased Monte Carlo estimator of the true variance of $\hat{\theta}$ (cf. Booth et al. 1994, p. 1287f).

Kuk (1989) tried to extend the idea of replicating the sample elements according to their design weights to generate a bootstrap population as the basis for the resampling process to systematic πPS sampling from a population ordered according to a size variable x providing an implicit stratification with respect to x. Therein, the actual replication factor for sample unit k is calculated by $c \cdot \frac{1}{\pi_k}$ assuming that this product is an integer. The constant c is chosen in a way that the bootstrap population U^* consists of $N^* = N$ units. For a pseudo-population generated in this way, it is shown for this specific application of the πPS sampling technique that the resampling procedure fails in estimating the variance of the HT estimator.

For this special sampling method, which is used widely in official statistics or in institutional surveys such as the PISA survey, Kuk (1989) proposes a method to generate a pseudo-population of original size N, in which the y-values of all units are estimated by a model relating y and x, where the values of x are known for all population units. The model is estimated by the sample data. With a residual component randomly chosen from these data, C different pseudo-populations are generated. The resampling is conducted within each of these pseudo-populations according to the original sampling plan. For C that is large enough, the bootstrap variance estimator (5.1) over all of these resamples approximates the variance of any statistic under this specific sampling scheme (cf. Kuk 1989, p. 75f).

Eventually, Holmberg (1998) developed a bootstrap approach to πPS sampling, or general π sampling, respectively (cf. Sect. 2.4.1). The design weight $\frac{1}{\pi_k} = \frac{t(x)}{x_k n}$ of survey unit $k \in U$ is decomposed into its integer part i_k and the "rest" r_k ($\frac{1}{\pi_k} = i_k + r_k$). To generate the pseudo-population U^*, each sample unit k is replicated i_k times and independently from each other, randomly once more with probability r_k. This process creates a pseudo-population of expected size N. Only for the unusual case that $r_k = 0$ applies for all sample elements, the bootstrap population U^* corresponds to the pseudo-population $U^*_{\pi PS}$ from πPS sampling (see Sect. 2.4.1).

After U^* is generated, the element bootstrap sample inclusion probabilities have to be recalculated according to the replicated variable x^* as a rule, resulting in $\pi_k^* = \frac{x_k^* n}{\sum_{U^*_{\pi PS}} x_k^*}$. Then, B πPS resamples of size n can be drawn from U^* according to the original πPS sampling scheme. After that, the estimation of parameter θ is done in the same way in each of these bootstrap samples as in the original one. For Pareto πPS sampling presented by Rosén (1997), it is shown in the study by Holmberg (1998) that for $N^* \to \infty$ and large n, the bootstrap variance estimator (5.1) achieves approximate unbiasedness with respect to the variance of the HT estimator of t (cf. Holmberg 1998, p. 381). With $\frac{x_k}{t^{(x)}} = \frac{1}{N}$ and $r_k = 0$ $\forall k \in U$, $U^* = U^*_{SI}$ applies and the approach by Holmberg reduces to that of Gross (1980) for SI schemes (see Fig. 5.1).

Barbiero and Mecatti (2010) aimed to simplify the procedure for πPS sampling suggested by Holmberg (1998) and, at the same time, improve its efficiency with respect to the estimation of the variance of $t_{\pi PS}$. They proposed to make "a more complete use of the auxiliary information" (Barbiero and Mecatti 2010, p. 62) available for an auxiliary variable x, in particular, for its total $t^{(x)}$. According to these authors, the following understandable properties should apply to a bootstrap algorithm with respect to the estimation of a total t of variable y (cf. Barbiero and Mecatti 2010, p. 60ff):

1. Given the sample s, in a bootstrap population U^*, the total $\sum_{U^*} x_k^*$ of the replicated values of an auxiliary variable x should be equal to the total $\sum_U x_k$ of x in the original population U.
2. Given s, the total $\sum_{U^*} y_k^*$ of the replications of the study variable y in U^* should be equal to $t_{\pi PS}$, the HT estimator of t.
3. Over the resampling process, for given s, the HT estimators of t calculated in the B resamples should have an expectation of $t_{\pi PS}$.

Obviously, these properties are desirable for an efficient estimation of the variance $V(t_{\pi PS})$ of the HT estimator $t_{\pi PS}$ according to Eq. (2.10) by $\hat{V}_{boot}(t_{\pi PS})$ according to Eq. (5.1). For different bootstrap methods dealing with the generation of bootstrap populations as described above, these properties only apply for $r_k = 0$ for all $k \in s$. For this reason, an "x-balanced πPS-bootstrap" is proposed by Barbiero and Mecatti (2010), where after replicating each sample unit k i_k times, further units are iteratively added to the bootstrap population up to the element, where the minimum difference of $\sum_{U^*} x_k^*$, when compared with the given parameter $t^{(x)}$ of the original population U, is achieved. In this process, the additional selection of units may simply start with that element k of s having the highest r_k-value or the highest ratio $q_k = \frac{1}{\pi_k} \cdot \frac{1}{i_k+1}$. Then, the element with the second highest rest-value r_k or ratio q_k is selected, and so on. In the second case, considering the q_k-values, for the same r_k-values, elements with a higher integer part i_k of the design weight $\frac{1}{\pi_k}$ have a higher probability of again being added to U^*, when compared with elements with a lower integer part. However, this procedure results in a pseudo-population with the total of the replicated x-values being close to the total of x in U. Note that by using $x = 1$ as an auxiliary variable, the method corresponds to Holmberg (1998) approach to the πPS-bootstrap for SI sampling with a pseudo-population $U^* = U^*_{SI}$ of fixed size $N^* = N$. At the end of the process, it would be more than natural that when more than one survey unit has the same r_k- or q_k-value, elements with x_k-values leading to an approximation closer to $t^{(x)}$ would be preferred. If such elements also have the same x_k-values, a random selection of the last element would complete the procedure.

After U^* is generated, the first-order inclusion probabilities π_k have to be recalculated by $\pi^*_k = \frac{x^*_k n}{\sum_{U^*} x^*_k}$ before the resampling process can start. Generating a bootstrap population that is close to the original population with respect to $t^{(x)}$ promises an improvement in the bootstrap estimation of the variance of the estimator under study (cf. Barbiero and Mecatti 2010, p. 63ff). But, also these proposals will not guarantee a size $N^* = N$ even for SI sampling, when $\frac{1}{\pi_k} \notin \mathbb{N}$ (cf. Ranalli and Mecatti 2012, p. 4095).

Another challenge for the bootstrap method is the incorporation of data imputation (see Sect. 3.3). When such a technique is applied to compensate for the nonresponse that occurred, treating the imputed values as if they were true ones will certainly underestimate the real inaccuracy of an estimator using imputed values. The reason is that this approach does not consider the uncertainty added by the imputations. This also applies to the bootstrap method of estimating such variances by Eq. (5.1). For such cases, Shao and Tu (1995) and Shao and Sitter (1996) proposed to generate a bootstrap population U^* of the same structure as U by replicating values of the imputed sample s_I (see Fig. 5.2). Therefore, U^* consists of replications y^* of the true y-values in the response set s_r and the imputed y-values in the missing set s_m (see Fig. 3.3). Then, as usual, B resamples $s^*_1, \ldots, s^*_B$ are drawn from the bootstrap population. In each of these B resamples from U^*, all values originally belonging to the missing set s_m of s are set to missing again. For these missings in s^*_b, the values are re-imputed by applying the same imputation method that was used in the original sample s. Information on auxiliary variables **x** available for all units in s^*_b is used for this purpose. This leads to B imputed bootstrap samples $s^*_{I,1}, \ldots, s^*_{I,B}$.

In each of these resamples, the estimator $\hat{\theta}_I$ of the parameter θ under study is calculated using the observed and imputed values. This is done by the HT approach of generating pseudo-populations $U^*_{I,b}$ from each imputed bootstrap sample $s^*_{I,b}$ (see Sect. 2.3). The distribution of these estimators $\hat{\theta}^*_{I,1}, \ldots, \hat{\theta}^*_{I,B}$ approximates the distribution of the imputed estimator $\hat{\theta}_I$ in the imputed sample s_I (see, for example,

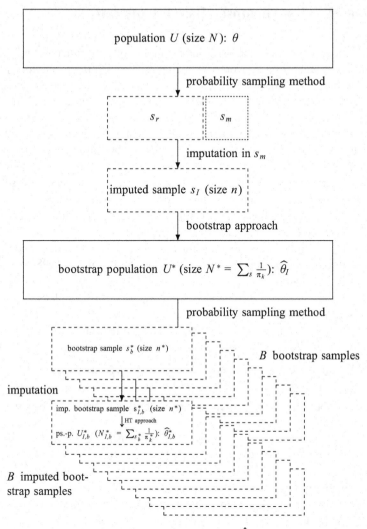

Fig. 5.2 Estimating the sampling distribution of an estimator $\hat{\theta}_I$ applying the bootstrap method in finite population sampling with data imputation

estimator t_I according to (3.4) from Sect. 3.3). This bootstrap technique incorporates the additional inaccuracy resulting from the use of imputed instead of true values and delivers asymptotically unbiased variance estimators without the use of a standard variance estimator (cf. Shao and Tu 1995; Shao and Sitter 1996, p. 1279ff).

5.3 The Horvitz–Thompson Based Bootstrap

The different approaches to the variance estimation of an estimator $\hat{\theta}$ for a finite population parameter θ by the bootstrap methods discussed in the previous section can be summarized as follows (see Fig. 5.2 for $s_m = \emptyset$): The process starts from the probability sample s, drawn according to a without-replacement probability sampling scheme with inclusion probabilities π_k ($k \in U$). For non-integer design weights, the presented bootstrap methods differ when it comes to the generation of the bootstrap population U^* of size N^* (or of different bootstrap populations). Such a bootstrap population U^* shall serve as set-valued estimator of U, particularly for simulations with respect to the variance of $\hat{\theta}$ under the given sampling scheme. Evidently, to mimic the sampling distribution of $\hat{\theta}$ in this respect, bootstrap population U^* must have the same structure regarding strata and/or clusters as the original population U (cf. Chao and Lo 1994, p. 398ff).

At this stage of the process, possibly, the π_k-values also have to be adapted to the original sampling method applied (cf., as an example, Holmberg 1998, p. 380). In the next step, B bootstrap samples $s_1^*, \ldots, s_B^*$ of size n^* are drawn from U^* with the original sampling method using these adapted inclusion probabilities π_k^*. In each resample s_b^*, the estimate $\hat{\theta}_b^*$ is calculated in the same way $\hat{\theta}$ was calculated in the original sample s. For this purpose, following the HT rule from Sect. 2.2 (Fig. 2.2), a pseudo-population U_b^* is generated by replicating the observed values of s_b^* $\frac{1}{\pi_k^*}$ times ($k \in s_b^*$). Within these B pseudo-populations $U_1^*, \ldots, U_B^*$, the estimates $\hat{\theta}_1^*, \ldots, \hat{\theta}_B^*$ are calculated. The distribution of these B $\hat{\theta}_b^*$-values serves as an estimator of the sampling distribution of $\hat{\theta}$. In particular, this means that its variance is estimated by (5.1).

For $U^* = U$, this bootstrap framework, which also includes Efron's original i.i.d. bootstrap as a special case, would perfectly simulate the sample-to-sample distribution of $\hat{\theta}$, which was shown in Fig. 2.1 (see Sect. 4). This is the intuitive approach of this type of finite population bootstrap. Therefore, the crucial point is the generation of the bootstrap population U^* as an estimator of U so that resamples can be drawn from U^* to mimic the actual sampling distribution of $\hat{\theta}$. As proposed above, when the parameter of interest is a function of population totals, the generation of U^* may directly follow the HT approach presented in Fig. 2.2 to estimate the original population U.

All methods described so far for the finite population bootstrap in πPS (or general π) sampling with non-integer design weights try to establish a bootstrap population to start the resampling process from it, which includes solely integer numbers of replications of the original sampling units, thus violating "the mimicking principle" (Ranalli and Mecatti 2012, p. 4095) of Efron's original bootstrap approach for non-integer design weights to an unknown extent. Subsequently, a procedure is presented, which is a natural development of the generation of a pseudo-population for the HT estimation of a total, as shown in Fig. 2.2 (Quatember

2014b). It complements the proposals of Holmberg (1998) and Barbiero and Mecatti (2010) for the problem of non-integer design weights in the most natural way.

In particular, this Horvitz–Thompson based bootstrap approach (HTB) allows also non-integer numbers of replications of the sample values of variables y and x to generate the bootstrap population U^*. This is done in the same way as in the HT estimator of a total to generate the HT pseudo-population U_{HT}^* (see Sect. 2.2). For a πPS sample, for example, let each unit k be replicated exactly $\frac{1}{\pi_k} = \frac{t^{(x)}}{x_k n}$ times. In this way, a bootstrap population $U^* = U_{\pi PS}^*$ is generated, which contains not only i_k whole units with values y_k and x_k, but also an additional r_k-piece of a unit with these values as it was described for the HT estimator in Sect. 2.2, when $r_k > 0$ applies ($k \in s$). This pseudo-population has the expected size $E(N^*) = E\left(\sum_s \frac{1}{\pi_k}\right) = N$. But, for SI sampling with $\pi_k = \frac{n}{N}$, which corresponds to πPS sampling with $x_k = 1$ $\forall \ k \in U$, for instance, this means that a bootstrap population $U^* = U_{SI}^*$ with size $N_{SI}^* = N$ is guaranteed. Note that the bootstrap population has to be generated in a way that the original sample s could also be a possible outcome of the sample selection process.

For $r_k > 0$ and increasing n, the difference between the standard deviation of size N^* of the HTB method and the finite population bootstrap methods allowing only integer numbers of replications of y-values from s increases. Furthermore, the difference between these standard deviations increases with less differing original first-order inclusion probabilities.

In the resampling process, based on this bootstrap population U^*, a whole unit k belonging to this population has a resample inclusion probability proportional to its original x-value. But, for an r_k-piece of a unit k, this probability is proportional to r_k times x_k. Hence, after the generation of U^* as a set-valued estimator of U, the design weights of the elements in U^* will not have to be recalculated. In each of the resamples drawn, the original estimator of the parameter under study is calculated.

For a πPS sample with $x_k = 1 \ \forall \ k \in U$ and $\frac{N}{n} \in \mathbb{N}$, the method reduces to the strategy for the SI technique proposed by Gross (1980) as discussed above. For arbitrary π_k-values and integer design weights, this procedure reduces to the techniques proposed by Holmberg (1998) and Barbiero and Mecatti (2010).

For the proposed HTB technique and a given without-replacement probability sample s, in the light of the three desirable properties mentioned above for efficient variance estimation (cf. Barbiero and Mecatti 2010, pp. 60ff), the following applies:

1. The total $\sum_{U^*} x_k^*$ of the replications of size variable x in U^* is given by: $\sum_{U^*} x_k^* = \sum_s x_k \cdot \frac{1}{\pi_k} = t^{(x)}$.
2. For the total $\sum_{U^*} y_k^*$ of the replications of variable y in U^*, $\sum_{U^*} y_k^* = \sum_s y_k \cdot \frac{1}{\pi_k} = t_{\pi PS}$ applies.
3. The expected value of the HT estimator of the total $\sum_{U^*} y_k^*$ of the replications y^* in U^* yields $E^*\left(\sum_{s_b^*} y_k^* \cdot \frac{1}{\pi_k}\right) = \sum_{U^*} y_k^* = t_{\pi PS}$ with E^* denoting the expectation over all resamples, given s and the sampling design.

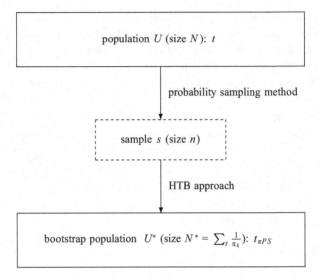

Fig. 5.3 Generating a bootstrap population for the HTB bootstrap method applying the HT approach of generating a pseudo-population

Nevertheless, for usual design weights, the proposed HTB method to generate the bootstrap population is not expected to perform considerably better than, for instance, the technique of Holmberg (1998). But, it might seem more understandable in terms of educational reasons than the more or less heuristic methods from literature presented in the previous section, because it follows the same idea as the one behind the widely used HT estimator when it comes to the composition of the bootstrap population (see Fig. 5.3). Additionally, the HTB bootstrap can still be used in πPS situations, where other methods fail because of first-order sample inclusion probabilities π_k of the population units which are close to one, because with the HTB method, these probabilities need not be recalculated before the resampling process.

5.4   An Example of the Application of the Finite Population Bootstrap

At the end of Sect. 2.7.2 on the capture–recapture (CR) method to estimate the population size, it was emphasized that, although variance estimators for the ratio estimator $N_{\text{rat(CR)}}$ (2.53) or the less-biased estimator by Chapman (1951) of the size N of population U exist, the usual approximate confidence interval based on the normal assumption of the estimator will hardly be valid, at least in small populations and samples. The reason is that the distributions of these size estimators are skewed to the right. Hence, confidence intervals based on the normal assumption do not cover the targeted confidence level $1 - \alpha$.

For such cases, estimating the actual distribution of a statistic such as $N_{rat(CR)}$ by the bootstrap method delivers not only an estimate of the theoretical variance, but also an alternative for the determination of confidence intervals which does not depend on assumptions regarding the sample distribution of the estimator. The percentile method for the construction of a reasonable $(1 - \alpha) \cdot 100\,\%$-confidence interval for a parameter θ directly uses the $(\alpha/2)$- and $(1 - \alpha/2)$-quantile of the observed bootstrap distribution of the estimator $\hat{\theta}$ as the lower and the upper bound of the confidence interval, respectively (cf. Efron 1981).

Again, the key question for the application of the bootstrap method to this finite population problem by generating a bootstrap population U^* is: How can an adequate bootstrap population (see, for example, Fig. 5.1), which may serve as the basis for the resampling process, be generated from the observed sample s? In the case of the CR method with the ratio estimator (2.53), a set-valued estimator of the recapture-ready population U_{CR} of unknown size N is the pseudo-population $U^* = U^*_{rat(CR)}$ generated from the original sample s by the ratio-corrected HT approach described in Sect. 2.7.2 (see Fig. 2.10).

To start with the drawing of the B bootstrap samples, the estimated size $N^* = N_{rat(CR)} = \sum_s \frac{1}{\pi_k} \cdot \frac{C}{\sum_s x_k \frac{1}{\pi_k}}$ of U^* can be rounded out to the nearest integer, resulting in a pseudo-population of size $[N_{rat(CR)}]$ ($[x]$ denotes the result of rounding out $x \in \mathbb{R}$ to the nearest integer). Remember that in Sect. 2.7.2, the auxiliary variable x indicates whether an observed sample unit is marked or not.

However, following the HTB approach to generate a bootstrap population by the general HT approach to survey sampling (see Sect. 5.3), the size of the bootstrap population does not have to be rounded out. Each sample element of s is replicated $\frac{1}{\pi_k} \cdot \frac{C}{\sum_s x_k \frac{1}{\pi_k}}$ times. This results in a bootstrap population of size $N^* = N_{rat(CR)}$ with C marked and $N^* - C$ unmarked elements. After that, the bootstrap process can start considering the given structure of population U with respect to the original sampling method applied (see Sect. 2.7.2).

For example, for the SI method, which is often used in this context at least as a model, the number $\sum_{s_b^*} x_k$ of marked elements in the resample s_b^* of size n, in principle, follows a hypergeometric distribution with parameters N^* and C (cf. Buckland 1984, p. 815f), but with N^* not necessarily being an integer. Hence, the first element of a resample s_b^* is a marked one with probability $\frac{C}{N^*}$ and an unmarked one with the remaining probability. Depending on the outcome of the first draw, the second element is a marked one with probability $\frac{C-1}{N^*-1}$ or $\frac{C}{N^*-1}$ and an unmarked one with the adequate remaining probabilities, and so on. In this way, n elements are drawn for the bth resample and a marked element has a probability

$$\frac{C - \sum_{s_{b_{j-1}}^*} h_{j-1}}{N^* - j + 1} \tag{5.2}$$

of being selected at the jth step of a draw by draw process to select the units for a resample from N^* ($j = 1, \ldots, n$). Therein, h_{j-1} denotes the number of times a marked element was already selected in the first $j-1$ steps of the process to generate

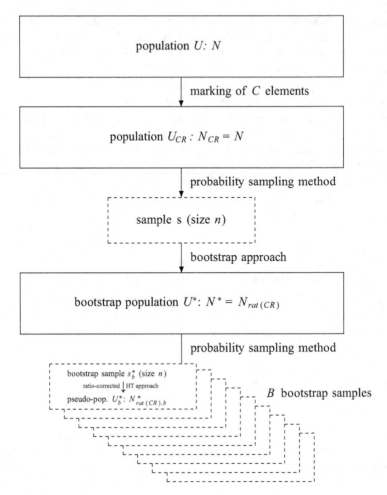

Fig. 5.4 Generating a bootstrap population for the application of the HTB bootstrap method with the capture–recapture (CR) technique

a resample ($h_0 = 0$). Furthermore, $s^*_{b_{j-1}}$ denotes the subset of the resample s^*_b after the $(j - 1)$th of n draws. The probability of the selection of an unmarked unit at the jth step of the process of drawing n resampling units is given by the complementary probability.

Altogether, B resamples are drawn from U^* following the original sampling scheme. In each one, the ratio estimator (2.53) is calculated. This gives a bootstrap estimation of the sample distribution of the CR estimator $N_{rat(CR)}$ of the unknown size N of the original population U (see Fig. 5.4). For large B, the ($\alpha/2$)- and ($1 - \alpha/2$)-quantiles of this empirical bootstrap distribution serve as reasonable bounds of an approximate confidence interval for the true population size N of the original population U with coverage probability $1 - \alpha$.

Chapter 6
Generalized Randomized Response Questioning Designs

6.1 Introduction

When questions on sensitive subjects, such as harassment at work, domestic violence, illegal employment, number of abortions, income, or voting behavior, are asked by direct questioning, nonresponse and untruthful answering will occur. As can be seen from Eq. (3.1), in the presence of both, the HT estimator t_{HT}, for instance, is decomposed into three sums: one over the truthful answering set s_t of sample s, another over the untruthful answering set s_u, and a third over the missing set s_m. Hence, such behavior by a respondent may cause serious problems in the analysis of sample and population data because the estimators of population parameters based only on a survey's available cases may strongly be biased. It is therefore essential for data collectors to not ignore nonresponse or untruthful answering. Before applying such methods as weighting adjustment and data imputation (see Sects. 3.2 and 3.3) to compensate for nonresponse that has already occurred, data collectors should do everything to make the rates of both nonresponse and untruthful answering as small as possible.

How can statistical science be embedded within the system of methods discussed by empirical social researchers, as presented in Sect. 3.1, to ensure a high level of data quality and quantity? For sensitive variables, randomized response (RR) questioning designs applied at the survey's design stage aim to address this question. A common characteristic of these methods is that, instead of directly asking for the sensitive variable, the question actually asked is randomly selected according to reasonably determined "design probabilities." The idea behind this questioning approach can be described as setting the variable y under study as missing in the whole sample and imputing the values y_k^i instead of y_k ($\forall\, k \in s$). This means that an RR design does not allow the data collector to assign the given answer of any survey element directly to the sensitive variable. This should reduce the individual's fear of an embarrassing "outing" and thus ensure that the responding person is willing to cooperate.

© Springer International Publishing Switzerland 2015
A. Quatember, *Pseudo-Populations*, DOI 10.1007/978-3-319-11785-0_6

The development of such techniques commenced with the pioneering work of Warner (1965) for the estimation of the relative size of a certain subgroup of the population. In this questioning design each respondent has to answer randomly either with probability p_I the question "Are you a member of this subgroup?" or with probability $p_{II} = 1 - p_I$ the alternative "Are you a member of the complementary subgroup?" $(0 < p_I < 1)$. Since then, various RR methods have been developed, for instance, to increase the efficiency of strategies to estimate proportions (examples of recent developments are Christofides 2003; Singh and Sedory 2011), to apply the idea to categorical and quantitative variables (see Sects. 6.2.1 and 6.3) or to include auxiliary information either at the survey's design or estimation stage (Diana and Perri 2009; Ryu et al. 2005). Other indirect questioning designs with no randomization device have been published (Groenitz 2014; Tan et al. 2009). Chaudhuri (2011) and Chaudhuri and Christofides (2013) wrote recent books on the theoretical aspects of such procedures. The practical use of these questioning designs is well documented (cf. the various papers on empirical sociological or psychological research quoted in Lensvelt-Mulders et al. 2005 applying the RR technique). The positive effect of these strategies on the response and the truthful answering rate, as well as on the perceived privacy protection when asking for sensitive information, has been repeatedly confirmed (see, for instance, the meta-analysis in Lensvelt-Mulders et al. 2005). It shall not be concealed that in some studies this positive effect was not found (cf., for instance, Holbrook and Krosnick 2010).

Warner (1971) was the first to note that the techniques of RR are also applicable as methods for statistical disclosure control (see Sect. 7) to "mask" confidential microdata sets in order to allow their release for public use (cf. Warner 1971, p. 887). When RR techniques are used in this context, either the survey units already perform data masking at the survey's design stage or the statistical agency applies the probability mechanism of the technique before the release of the microdata file (cf., for instance, van den Hout and van der Heijden 2002).

6.2 The Estimation of the Sizes of Disjointed Subgroups by the Randomized Response Technique

6.2.1 A Generalized Questioning Design for Categorical Variables

After Warner's kick-start, more basic ideas were presented, for instance, by Horvitz et al. (1967) (with its theoretical framework presented by Greenberg et al. 1969) and Boruch (1971). Other studies considered chaining and combining such proposals. The estimators found in this way were apparently different from each other. The efficiency comparisons rarely, hardly ever in fact, considered the privacy protection offered by the different techniques although "since the degree of privacy is an essential part of the randomized response procedure, and greater privacy will, in general, have greater costs in terms of variance of the estimate, one obvious basis for comparing randomized models is to compare variances only when the required degree of privacy is held constant" (Warner 1976, p. 205). Often, efficiency comparisons were done by determining those randomization probabilities that provide a better performance for the new strategy compared to others, as if the choice of these design probabilities could not affect the perceived privacy protection and therefore the respondents' willingness to cooperate. Despite attempts to develop a general framework (cf., for instance, Chaudhuri 2001; Quatember 2009), the main result over the years has been to inflate the respective literature with a surplus supply of theory "not keeping pace in practice" (Chaudhuri 2011, p. xiii).

Motivated by these facts, a unified approach for the estimation of the category sizes of a categorical variable is introduced in the following. It brings together the work of Quatember (2014a), which was based on Quatember (2009) and Quatember (2012). Moreover, the theory is extended to the case that for some of the respondents their true values of the sensitive variable can be observed.

This framework is broad enough to encompass as special cases several techniques already published and the rest still "awaiting" their individual presentations. Furthermore, whereas "almost invariably, the randomized response theory in the literature till date is apparently connected to the case of simple random sampling with replacement" (see Chaudhuri 2011, p. 1), the theory of the strategy proposed in the following is developed for without-replacement probability sampling schemes (see Sect. 2.4) with arbitrary first-order inclusion probabilities π_k ($k = 1, \ldots, N$). This seems to be important for the practical applicability of such theoretical ideas because in fields, in which sensitive questions are asked, complex sampling schemes with differing first-order sample inclusion probabilities, including stratification (by region or sex, for example) or clustering (such as in household surveys) are often used.

The unified questioning design can be formulated in the following way: Let universe U of N population units be divided by a categorical variable y with categories $1, \ldots, H$ into $H \geq 2$ non-overlapping subgroups $U_1, \ldots, U_H$ of sizes $N_1, \ldots, N_H$ ($U = \bigcup U_h$, $N = \sum N_h$, $U_h \cap U_i = \emptyset \; \forall \; h \neq i; h, i = 1, \ldots, H$). Furthermore, let the parameters of interest be the sizes $N_1, \ldots, N_H$ of the H subgroups. For instance, one might be interested in the number of elements of certain subgroups itself, or in the case of an ordinal or a quantitative discrete variable y, proportions may be needed to calculate measures of position, dispersion, or other distribution characteristics.

For element k, let the value y_k be the true category i of variable y ($y_k = i$). Moreover,

$$y_{kh} = \begin{cases} 1 & \text{if } y_k = h, \\ 0 & \text{otherwise} \end{cases}$$

indicates the membership of group U_h ($h = 1, \ldots, H$). The total of the y_{kh}-values over all population units $k = 1, \ldots, N$ for category h is $N_h = \sum_U y_{kh}$.

Let s be a without-replacement probability sample of size n. Aiming to increase the survey units' perception of privacy protection to reduce nonresponse and untruthful answering, a generalized RR questioning design R can be formulated in the following way: A drawn sample unit k is asked

- with probability p_I for his or her true value $y_k = i$ of variable y (Instruction I);
- with probability p_{II} to choose randomly one of the $H - 1$ categories $h \neq i$ of y that he or she does not belong to according to preassigned probabilities ($h = 1, \ldots, H, h \neq i$) (Instruction II);
- with probability p_{III} to choose randomly one of all H categories $1, \ldots, H$ of y according to preassigned probabilities (Instruction III); or
- with probability p_{IV} for his or her true value $x_k = j$ of a variable x with H categories ($j = 1, \ldots, H$) (Instruction IV).

The sum of the probabilities p_I, p_{II}, p_{III}, and p_{IV} equals 1. The probability p_I of being asked the direct question on variable y can be composed of the design probabilities from, say, $K \geq 1$ stages in which the direct question might be asked (cf. Quatember 2012). The process might start at stage 1, at which Instruction I is presented with probability $p_I^{(1)} < 1$. With the remaining probability, $1 - p_I^{(1)}$, the respondent is directed to a second stage, at which this question is asked with probability $p_I^{(2)}$. With the remaining probability, the survey unit might be directed to a third stage and so on. At the final Kth stage, the survey unit is asked the question on y with probability $p_I^{(K)}$. With the probabilities p_{II}, p_{III}, and p_{IV}, Instructions II–IV are given to the survey unit at the final stage. This gives a total probability

$$p_I = p_I^{(1)} + \left(1 - p_I^{(1)}\right) \cdot p_I^{(2)} + \cdots + \left(1 - p_I^{(1)}\right) \cdot \left(1 - p_I^{(2)}\right) \cdot \ldots \cdot \left(1 - p_I^{(K-1)}\right) \cdot p_I^{(K)}$$

of being asked the question on the variable y under study (Instruction I). Some authors of RR techniques have suggested such multi-stage questioning designs instead of a one-stage version with $p_I = p_I^{(1)}$, apparently to increase the privacy protection as perceived subjectively by the respondents at the cost of a more complex procedure (cf., for instance, Bourke and Dalenius 1976, p. 220).

The probability p_{III} to get the instruction to choose randomly one of the H categories is decomposed into a sum of H probabilities $p_{III,1}, \ldots, p_{III,H}$ ($\sum_h p_{III,h} = p_{III}$). These can be calculated from the preassigned conditional probabilities $p_{1|III}, \ldots, p_{H|III}$ of choosing categories 1 to H if Instruction III is selected ($\sum_h p_{h|III} = 1$), resulting in $p_{III,h} = p_{h|III} \cdot p_{III}$ ($h = 1, \ldots, H$). The not necessarily equal probabilities $p_{h|III}$ can be determined by the data collector in advance according to, for instance, different sensitivity levels of the membership of certain categories. When it comes to the allocation of Instruction II to a survey unit, one could change these probabilities. However because they should be reasonably assigned, there is a non-necessity to change them. Using the same category probabilities as for Instruction III results in $p_{\mathrm{II},h} = \frac{p_{h|\mathrm{III}}}{\sum_{h \neq i} p_{h|\mathrm{III}}} \cdot p_{\mathrm{II}}$ ($h = 1, \ldots, H$; $h \neq i$). For equal conditional probabilities, $p_{h|\mathrm{III}} = \frac{1}{H}$, $p_{\mathrm{III},h} = \frac{p_{\mathrm{III}}}{H}$, and $p_{\mathrm{II},h} = \frac{p_{\mathrm{II}}}{H-1}$ apply.

As another alternative to the question on the sensitive variable y, in the generalized randomization technique R according to the idea by Horvitz et al. (1967), a survey unit is asked with probability p_{IV} for the value x_k of a completely nonsensitive variable x not related to y and having the same number H of categories as y, denoted as $1, \ldots, H$. For this purpose, the category sizes $N_h^{(x)}$ of category h of x must be known for the population so that $p_{\mathrm{IV},h} = \frac{N_h^{(x)}}{N} \cdot p_{\mathrm{IV}}$ and $\sum_h p_{\mathrm{IV},h} = p_{\mathrm{IV}}$ apply.

The difference between the random selection of a category h with probabilities $p_{\mathrm{III},h}$ and $p_{\mathrm{IV},h}$, respectively, lies in the random nature of the selection. In the first case, the answer of a certain survey unit k can take different values, whereas in the second case, the random nature stems solely from the grouping of U according to variable x. In Instruction IV, the answer category h is a fixed value for a given survey unit k similar to variable y in Instruction I.

Clearly, for $p_{\mathrm{I}} < 1$, the questioning design R does not enable the data collector to identify with certainty the randomly selected question or instruction; therefore, it protects the respondent's privacy. Practicable randomizing devices for this procedure can be found in different sources (see, for instance, Warner 1986 for telephone surveys, and Boeije and Lensvelt-Mulders 2002 or Peeters et al. 2010, p. 290ff, for computer-assisted RR).

All possible combinations of the four selectable instructions defined above shall be included in the unified approach R under one theoretical umbrella. In the following, some of these combinations, which to the author's knowledge have already been published are mentioned. The list starts with the direct questioning design:

- For $p_{\mathrm{I}} = 1$, the direct questioning on the subject is included in strategy R.

For $H = 2$ categories, the following different ideas are special cases of R, in all of which $0 < p_{\mathrm{I}} < 1$ applies and the design probabilities $p_{\mathrm{II}}, p_{\mathrm{III}}$, and p_{IV} not mentioned explicitly are set to zero:

- For $p_{\mathrm{II}} = 1 - p_{\mathrm{I}}$, Warner (1965) and its two-stage version by Mangat and Singh (1990) are members of family R.
- With $p_{\mathrm{III},1}, p_{\mathrm{III},2} > 0$ and $p_{\mathrm{III}} = 1 - p_{\mathrm{I}}$, the technique of Boruch (1971) and its two-stage version by Singh et al. (1995) belong to our framework R.
- The choice of $p_{\mathrm{III},1} = p_{\mathrm{III}}$ for $p_{\mathrm{III}} = 1 - p_{\mathrm{I}}$ results in an RR questioning design applied by Quatember (2009).
- With $p_{\mathrm{IV}} = 1 - p_{\mathrm{I}}$, Horvitz et al. (1967) and its two-stage version by Mangat (1992) are incorporated into R.
- With $p_{\mathrm{II}} > 0$, $p_{\mathrm{III},2} = p_{\mathrm{III}}$, and $p_{\mathrm{II}} + p_{\mathrm{III}} = 1 - p_{\mathrm{I}}$, Mangat et al. (1995) and its two-stage version previously published by Mangat et al. (1993) become special cases of the unified framework R.

- For $p_{II} > 0$, $p_{III,1} = p_{III}$, and $p_{II} + p_{III} = 1 - p_I$, the RR family R also includes the idea of Bhargava and Singh (2000).
- Fixing design probabilities as $p_{II} > 0$, $p_{III,1}, p_{III,2} > 0$ and $p_{II} + p_{III} = 1 - p_I$ results in the idea suggested by Chang et al. (2004).
- For $p_{II} > 0$, $p_{IV} > 0$, and $p_{II} + p_{IV} = 1 - p_I$, Bourke (1984) is also found under the unified theoretical umbrella R.
- Choosing $p_{III,1} = p_{III}$, $p_{IV} > 0$, and $p_{III} + p_{IV} = 1 - p_I$, Perri (2008) becomes a member of the family R.
- With $p_{III,2} = p_{III}$, $p_{IV} > 0$, and $p_{III} + p_{IV} = 1 - p_I$, Singh et al. (2003) and its two-stage version previously published by Singh et al. (1994) are members of the unified approach.
- For $p_{II} > 0$, $p_{III,1}, p_{III,2} > 0$, $p_{IV} > 0$, and $p_{II} + p_{III} + p_{IV} = 1 - p_I$ the standardized RR questioning design by Quatember (2009) and its multi-stage equivalent included in the family defined by Quatember (2012) are found under the same umbrella R.

Moreover, for $H > 2$ categories,

- with design probabilities $0 < p_I < 1$, $p_{III,h} > 0 \; \forall \, h = 1, \ldots, H$, and $p_{III} = 1 - p_I$, the generalized RR technique R results in the design by Liu and Chow (1976) as discussed in Quatember (2014a).

Just to assure that there can be no misunderstanding: This is not an invitation to use all four Instructions with design probabilities larger than zero. Rather, the objective is to provide a theoretical framework that can be applied with all combinations of these Instructions for $p_I, p_{II}, p_{III}, p_{IV} \geq 0$.

Now, let z_k be the response category l of survey unit k ($z_k = l$) with respect to the randomized response questioning design R and

$$z_{kh} = \begin{cases} 1 & \text{if } z_k = h, \\ 0 & \text{otherwise} \end{cases}$$

($h = 1, \ldots, H$). At this place, the idea of generating a pseudo-population offers the opportunity to discuss the theoretical properties of the questioning design R for general probability sampling instead of the limitation to i.i.d. sampling, which is assumed in the vast majority of the relevant literature. For this purpose assuming cooperation, the probability of $z_{kh} = 1$ with respect to R for given y_{kh} is calculated by

$$P_R(z_{kh} = 1) = p_I \cdot y_{kh} + p_{II,h} \cdot (1 - y_{kh}) + p_{III,h} + p_{IV,h}$$
$$= u_h + v_h \cdot y_{kh}$$

with $u_h \equiv p_{II,h} + p_{III,h} + p_{IV,h}$ and $v_h \equiv p_I - p_{II,h}$. Thus, the term

$$y_{kh}^i = \frac{z_{kh} - u_h}{v_h} \tag{6.1}$$

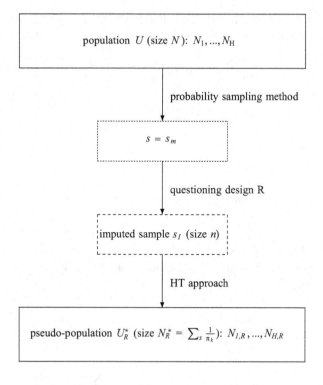

Fig. 6.1 Generating a pseudo-population with the questioning design R

is used as an unbiased estimator for the true value y_{kh} ($v_h \neq 0$). To be able to estimate the relative group sizes of variable y, the estimator (3.4) from data imputation can be applied considering the special case of $s_r = \emptyset$ and $s_m = s$ (see Fig. 6.1). The y_{kh}^i-values can be viewed as "imputations" of the unknown, missing y_{kh}-values. Adapting the derivations in Quatember (2012, 2014a) to the questioning design R results in the following theorems.

Theorem 1a *For a without-replacement probability sampling scheme P with arbitrary first-order inclusion probabilities π_k ($k \in U$) and the RR questioning design R,*

$$N_{h,R} = \sum_s y_{kh}^i \cdot \frac{1}{\pi_k} \tag{6.2}$$

is an unbiased estimator of size $N_h = \sum_U y_{kh}$ of group U_h in population U ($h = 1, \ldots, H$).

Proof With E_P and E_R, denoting the expectation over the probability sampling design P and over the randomized questioning design R, respectively,

$$E(N_{h,R}) = E_P \left[E_R \left(\sum_s y_{kh}^i \cdot \frac{1}{\pi_k} \Big| s \right) \right] = E_P \left(\sum_s y_{kh} \cdot \frac{1}{\pi_k} \right)$$

$$= \sum_U y_{kh} = N_h$$

applies.

Based on (6.2), we can describe the estimation of N_h as a special case of the imputation estimator t_I (3.4). After the sampling units are drawn with probability sampling method P, the questioning design R is applied to all sample elements of s delivering an "imputed sample" s_I of size n. Considering the sampling method, the HT approach for the estimation of a population total described in Sect. 2.2 results in the generation of a pseudo-population U_R^* of size $N_R^* = \sum_s \frac{1}{\pi_k}$ elements, with $\frac{1}{\pi_1}$ units having the imputed value y_{1h}^i of the replication variable y_h^*, $\frac{1}{\pi_2}$ units having the value y_{2h}^i, and so on, for each category h $(h = 1, \ldots, H)$. The estimators $N_{1,R}, \ldots N_{H,R}$ of the sizes $N_1, \ldots N_H$ of the H different subgroups $U_1, \ldots, U_H$ of U can be counted directly using the replicated variable values y_{kh}^{i*} of units k for category h in pseudo-population U_R^* (Fig. 6.1). Applying a presentation of type (2.9), (6.2) can be written as

$$N_{h,R} = \sum_{U_R^*} y_{kh}^{i*}. \tag{6.3}$$

Theorem 1b *The theoretical variance of $N_{h,R}$ is given by*

$$V(N_{h,R}) = \sum \sum_U \Delta_{kl} \cdot \frac{y_{kh}}{\pi_k} \cdot \frac{y_{lh}}{\pi_l} + a_h \cdot \sum_U \frac{1}{\pi_k} + (b_h - 1) \cdot \sum_U y_{kh} \cdot \frac{1}{\pi_k} \tag{6.4}$$

with $a_h \equiv \frac{u_h(1-u_h)}{v_h^2}$, $b_h \equiv \frac{1-2u_h}{v_h}$ $(h = 1, \ldots, H)$.

Proof The theoretical variance of estimator (6.2) can be calculated by

$$V(N_{h,R}) = V_P(E_R(N_{h,R}|s)) + E_P(V_R(N_{h,R}|s)),$$

with V_P and V_R denoting the variances over the probability sampling procedure P and questioning design R, respectively. For the first summand,

$$V_P(E_R(N_{h,R}|s)) = V_P \left(\sum_s y_{kh} \cdot \frac{1}{\pi_k} \right)$$

applies, which is given by $V(t_{HT})$ from Eq. (2.5). Let I_k indicate the sample inclusion of a population unit k (see Sect. 2.1) with $E_P(I_k) = \pi_k$. Further, performing the randomization procedure R independently on each individual results in a covariance of y_{kh}^i and y_{lh}^i of $C_R(y_{kh}^i, y_{lh}^i | s) = 0 \ \forall \ k \neq l$. Therefore, with $E_P(I_k^2) = E_P(I_k)$, the following expectation over P of the variances over R is derived:

$$E_P(V_R(N_{h,R}|s)) = E_P\left(V_R\left(\sum_U I_k \cdot y_{kh}^i \cdot \frac{1}{\pi_k} \middle| s\right)\right)$$

$$= E_P\left(\sum_U I_k^2 \cdot \frac{1}{\pi_k^2} \cdot V_R(y_{kh}^i)\right)$$

$$= \sum_U V_R(y_{kh}^i) \cdot \frac{1}{\pi_k}.$$

Variance $V_R(y_{kh}^i)$ is given by

$$V_R(y_{kh}^i) = \frac{1}{v_h^2} \cdot V_R(z_{kh})$$

and because $z_{kh}^2 = z_{kh}$ and $y_{kh}^2 = y_{kh}$,

$$V_R(z_{kh}) = E_R(z_{kh}^2) - E_R^2(z_{kh})$$

$$= v_h \cdot y_{kh} + u_h - (v_h \cdot y_{kh} + u_h)^2$$

$$= u_h \cdot (1 - u_h) + v_h \cdot (1 - v_h - 2 \cdot u_h) \cdot y_{kh}$$

applies. Hence,

$$E_P(V_R(N_{h,R}|s)) = \frac{1}{v_h^2} \cdot \left(u_h \cdot (1 - u_h) \cdot \sum_U \frac{1}{\pi_k} + \right.$$

$$\left. + v_h \cdot (1 - v_h - 2 \cdot u_h) \cdot \sum_U y_{kh} \cdot \frac{1}{\pi_k}\right).$$

This completes the proof of (6.4).

The first summand of (6.4) corresponds to the variance formula for the direct questioning under the assumption of full cooperation. The second one can be regarded as the price that has to be paid by the data analyst for the increased privacy protection of the respondents and the reduced risk of nonresponse. For $s = U$ (and $n = N$), (6.4) reduces to $a_h \cdot N + (b_h - 1) \cdot N_h$.

Theorem 1c *The theoretical variance $V(N_{h,R})$ according to (6.4) can be estimated unbiasedly by*

$$\hat{V}(N_{h,R}) = \sum\sum_s \frac{\Delta_{kl}}{\pi_{kl}} \cdot \frac{y_{kh}^i}{\pi_k} \cdot \frac{y_{lh}^i}{\pi_l} + a_h \cdot N_R^* + (b_h - 1) \cdot N_{h,R} \qquad (6.5)$$

$(h = 1, \ldots, H).$

Proof For the proof of the unbiasedness of (6.5) with regard to the theoretical variance $V(N_{h,R})$ of $N_{h,R}$, we begin with

$$V(N_{h,R}) = \sum\sum_U \Delta_{kl} \cdot \frac{y_{kh}}{\pi_k} \cdot \frac{y_{lh}}{\pi_l} + a_h \cdot \sum_U \frac{1}{\pi_k}$$

$$+ (b_h - 1) \cdot \sum_U y_{kh} \cdot \frac{1}{\pi_k}$$

$$= \sum\sum_U \Delta_{kl} \cdot \frac{y_{kh}}{\pi_k} \cdot \frac{y_{lh}}{\pi_l} + a_h \cdot N + (b_h - 1) \cdot N_h$$

$$+ a_h \cdot \sum_U \left(\frac{1}{\pi_k} - 1\right) + (b_h - 1) \cdot \sum_U y_{kh} \cdot \left(\frac{1}{\pi_k} - 1\right).$$

The expectation of $\sum\sum_s \frac{\Delta_{kl}}{\pi_{kl}} \cdot \frac{y_{kh}^i}{\pi_k} \cdot \frac{y_{lh}^i}{\pi_l}$ over the questioning design R is given by:

$$E_R \left(\sum\sum_s \frac{\Delta_{kl}}{\pi_{kl}} \cdot \frac{y_{kh}^i}{\pi_k} \cdot \frac{y_{lh}^i}{\pi_l}\right) = \sum\sum_U I_k \cdot I_l \cdot \frac{\Delta_{kl}}{\pi_{kl}} \cdot \frac{E_R(y_{kh}^i \cdot y_{lh}^i)}{\pi_k \cdot \pi_l}$$

$$= \sum\sum_{U(k \neq l)} I_k \cdot I_l \cdot \frac{\Delta_{kl}}{\pi_{kl}} \cdot \frac{y_{kh}}{\pi_k} \cdot \frac{y_{lh}}{\pi_l}$$

$$+ \sum_U I_k^2 \cdot \frac{\Delta_{kk}}{\pi_k} \cdot \frac{E_R[(y_{kh}^i)^2]}{\pi_k^2}.$$

Therein,

$$E_R[(y_{kh}^i)^2] = E_R \left(\frac{z_{kh}^2 - 2 \cdot u_h \cdot z_{kh} + u_h^2}{v_h^2}\right)$$

$$= \frac{1}{v_h^2} \cdot \left((1 - 2 \cdot u_h) \cdot E_R(z_{kh}) + u_h^2\right)$$

$$= \frac{(1 - 2 \cdot u_h) \cdot v_h \cdot y_{kh} + u_h \cdot (1 - u_h)}{v_h^2}$$

$$= b_h \cdot y_{kh} + a_h.$$

Next,

$$E_P \left(E_R \left(\sum\sum_s \frac{\Delta_{kl}}{\pi_{kl}} \cdot \frac{y_{kh}^i}{\pi_k} \cdot \frac{y_{lh}^i}{\pi_l} \middle| s\right)\right)$$

$$= \sum\sum_{U(k \neq l)} \frac{\Delta_{kl}}{\pi_{kl}} \cdot \frac{y_{kh}}{\pi_k} \cdot \frac{y_{lh}}{\pi_l} \cdot E_P(I_k \cdot I_l)$$

$$+ \sum_U \frac{\Delta_{kk}}{\pi_k} \cdot \frac{a_h + b_h \cdot y_{kh}}{\pi_k^2} \cdot E_P(I_k^2)$$

$$= \sum\sum_U \Delta_{kl} \cdot \frac{y_{kh}}{\pi_k} \cdot \frac{y_{lh}}{\pi_l} + \sum_U \frac{\Delta_{kk}}{\pi_k^2} \cdot (a_h + (b_h - 1) \cdot y_{kh})$$

$$= \sum\sum_U \Delta_{kl} \cdot \frac{y_{kh}}{\pi_k} \cdot \frac{y_{lh}}{\pi_l} + \sum_U (1 - \pi_k) \cdot (a_h + (b_h - 1) \cdot y_{kh}) \cdot \frac{1}{\pi_k}$$

$$= \sum\sum_U \Delta_{kl} \cdot \frac{y_{kh}}{\pi_k} \cdot \frac{y_{lh}}{\pi_l} + a_h \cdot \sum_U \left(\frac{1}{\pi_k} - 1 \right)$$

$$+ (b_h - 1) \cdot \sum_U y_{kh} \cdot \left(\frac{1}{\pi_k} - 1 \right).$$

With $E(N_R^*) = E(\sum_s \frac{1}{\pi_k}) = N$ and $E(N_{h,R}) = N_h$, Eq. (6.5) is unbiased for $V(N_{h,R})$.

Theorem 1d *The theoretical covariance* $C(N_{h,R}, N_{j,R})$ *of estimators* $N_{h,R}$ *and* $N_{j,R}$ *with* $h \neq j$ *is given by*

$$C(N_{h,R}, N_{j,R}) = -\frac{1}{v_h \cdot v_j} \cdot \sum_U (u_h \cdot v_j \cdot y_{kj} + u_j \cdot v_h \cdot y_{kh} + u_h \cdot u_j) \cdot \frac{1}{\pi_k}$$

$$+ \sum\sum_{U(k \neq l)} y_{kh} \cdot y_{lj} \cdot \frac{\pi_{kl}}{\pi_k \cdot \pi_l} - N_h \cdot N_j \qquad (6.6)$$

$(h \neq j; h, j = 1, 2, \ldots, H).$

Proof For the covariance applies

$$C(N_{h,R}, N_{j,R}) = E\left(\left(\sum_s y_{kh}^i \cdot \frac{1}{\pi_k} \right) \cdot \left(\sum_s y_{lj}^i \cdot \frac{1}{\pi_l} \right) \right) - N_h \cdot N_j$$

$$= E_P \left(E_R \left(\sum_s y_{kh}^i \cdot y_{kj}^i \cdot \frac{1}{\pi_k^2} \bigg| s \right) \right)$$

$$+ E_P \left(E_R \left(\sum\sum_{s(k \neq l)} y_{kh}^i \cdot y_{lj}^i \cdot \frac{1}{\pi_k} \cdot \frac{1}{\pi_l} \bigg| s \right) \right)$$

$$- N_h \cdot N_j.$$

In the first of the two summands, the expectation of the product $y_{kh}^i \cdot y_{kj}^i$ over the randomization strategy R is given by

$$E_R(y_{kh}^i \cdot y_{kj}^i | s) = E_R \left(\frac{z_{kh} - u_h}{v_h} \cdot \frac{z_{kj} - u_j}{v_j} \bigg| s \right)$$

$$= \frac{1}{v_h \cdot v_j} \cdot E_R(-u_h \cdot z_{kj} - u_j \cdot z_{kh} + u_h \cdot u_j | s)$$

$$= -\frac{1}{v_h \cdot v_j} \cdot (u_h \cdot v_j \cdot y_{kj} + u_j \cdot v_h \cdot y_{kh} + u_h \cdot u_j).$$

In the second summand, for the expectation over R with $k \neq l$

$$E_R(y_{kh}^i \cdot y_{lj}^i | s) = E_R(y_{kh}^i | s) \cdot E_R(y_{lj}^i | s) = y_{kh} \cdot y_{lj}$$

applies. This results in

$$C(N_{h,R}, N_{j,R})$$

$$= E_P\left(-\frac{1}{v_h \cdot v_j} \cdot \sum_U (u_h \cdot v_j \cdot y_{kj} + u_j \cdot v_h \cdot y_{kh} + u_h \cdot u_j) \cdot \frac{1}{\pi_k^2} \cdot I_k\right)$$

$$+ E_P\left(\sum\sum_{U(k \neq l)} y_{kh} \cdot y_{lj} \cdot \frac{1}{\pi_k} \cdot \frac{1}{\pi_l} \cdot I_k \cdot I_l\right) - N_h \cdot N_j.$$

Because $E_P(I_k) = \pi_k$ and $E_P(I_k \cdot I_l) = \pi_{kl}$ (see Sect. 2.1), the correctness of Theorem 1d is proven.

Theorem 1e *The covariance* (6.6) *is estimated unbiasedly by*

$$\hat{C}(N_{h,R}, N_{j,R}) = -\frac{1}{v_h \cdot v_j} \cdot \sum_s (u_h \cdot v_j \cdot y_{kj}^i + u_j \cdot v_h \cdot y_{kh}^i + u_h \cdot u_j) \cdot \frac{1}{\pi_k^2}$$

$$+ \sum\sum_{s(k \neq l)} \frac{\Delta_{kl}}{\pi_{kl}} \cdot \frac{y_{kh}^i}{\pi_k} \cdot \frac{y_{lj}^i}{\pi_l} \tag{6.7}$$

$(h \neq j; h, j = 1, 2, \ldots, H).$

Proof The following expectations have to be derived:

$$E_P\left(E_R\left(\sum_s y_{kh}^i \cdot \frac{1}{\pi_k^2} \Big| s\right)\right) = E_P\left(\sum_U y_{kh} \cdot \frac{1}{\pi_k^2} \cdot I_k\right) = \sum_U y_{kh} \cdot \frac{1}{\pi_k}$$

and

$$E_P\left(E_R\left(\sum_s \frac{1}{\pi_k^2} \Big| s\right)\right) = E_P\left(\sum_U \frac{1}{\pi_k^2} \cdot I_k\right) = \sum_U \frac{1}{\pi_k}.$$

With these results, the unbiasedness of the first summand of (6.7) for the first summand of (6.6) is immediately proven. With respect to the last two summands of (6.6), firstly

$$E_P\left(E_R\left(\sum\sum_{s(k \neq l)} y_{kh}^i \cdot y_{lj}^i \cdot \frac{1}{\pi_k \cdot \pi_l} \Big| s\right)\right)$$

$$= E_P\left(\sum\sum_{U(k \neq l)} y_{kh} \cdot y_{lj} \cdot \frac{1}{\pi_k \cdot \pi_l} \cdot I_k \cdot I_l\right)$$

$$= \sum\sum_{U(k \neq l)} y_{kh} \cdot y_{lj} \cdot \frac{\pi_{kl}}{\pi_k \pi_l}$$

applies. Furthermore, for the term $\sum \sum_{s(k \neq l)} y_{kh}^i \cdot y_{lj}^i \cdot \frac{1}{\pi_{kl}}$, the following expectation over P and R is derived:

$$
E_P \left(E_R \left(\sum \sum_{s(k \neq l)} y_{kh}^i \cdot y_{lj}^i \cdot \frac{1}{\pi_{kl}} \Big| s \right) \right)
$$

$$
= E_P \left(\sum \sum_{U(k \neq l)} y_{kh} \cdot y_{lj} \cdot \frac{1}{\pi_{kl}} \cdot I_k \cdot I_l \right)
$$

$$
= \sum \sum_{U(k \neq l)} y_{kh} \cdot y_{lj} = N_h \cdot N_j.
$$

The development

$$
\sum \sum_{s(k \neq l)} y_{kh}^i \cdot y_{lj}^i \cdot \left(\frac{1}{\pi_k \cdot \pi_l} - \frac{1}{\pi_{kl}} \right) = \sum \sum_{s(k \neq l)} y_{kh}^i \cdot y_{lj}^i \cdot \frac{\pi_{kl} - \pi_k \cdot \pi_l}{\pi_k \cdot \pi_l \cdot \pi_{kl}}
$$

completes the proof of the unbiasedness of (6.7) for (6.6).

Together with the variances, statistical hypotheses tests on the difference between two category sizes can be performed. Such an issue arises frequently in opinion polls. The expressions given in the five theorems are general for probability sampling and must be worked out for a specific sampling scheme using its specific sample inclusion probabilities.

6.2.2 The Estimation of the Size of Very Small or Very Large Categories

The estimators $N_{h,R}$ may be outside the interval $[0; N]$ if one or more of the respective category sizes N_h are very small or very large ($h = 1, \ldots, H$). It follows that the H estimators $N_{h,R}$ according to (6.2) are moment but not ML estimators of these parameters. For this reason, Mangat and Singh (1991) recommended sequentially continuing sampling until a predetermined number of certain responses are observed. These numbers are chosen in such a way that the probability for estimators to be smaller than zero or larger than one is small. As with all sequential schemes, the problem of unlikely, but possibly, large sample sizes limits the applicability of the proposed procedure.

To determine the ML estimators of the N_h-values instead of the moment estimators, the EM algorithm can be applied (cf. Dempster et al. 1977). For this purpose, the observed responses z have to be viewed as a mixture with mixing proportions $\frac{N_h}{N}$. The unobserved true variable y is completely missing (cf, for instance, Bourke and Moran 1988, p. 966). As applied in Quatember (2014a), at iteration t the E step of the EM algorithm replaces each missing value y_{kh} by its expected value, conditioned on z_{kj} and the estimate $N_{h,R}$ of the tth iteration. This expectation is given by

$$
E^{(t)}(y_k = h | z_k = j, N_{h,R}^{(t)}) = \frac{P(z_{kj} = 1 | y_{kh} = 1) \cdot N_{h,R}^{(t)}}{\sum_h P(z_{kj} = 1 | y_{kh} = 1) \cdot N_{h,R}^{(t)}} \tag{6.8}
$$

$(h, j = 1, \ldots, H)$. The M step of the EM algorithm maximizes the likelihood of the observations. This corresponds to an "update" $t+1$ of the estimator $N_{h,R}^{(t)}$ considering probability sampling procedure P:

$$N_{h,R}^{(t+1)} = \sum_{s} \left(\sum_{j} z_{kj} \cdot E^{(t)}(y_k = h | z_k = j, N_{h,R}^{(t)}) \right) \cdot \frac{1}{\pi_k}. \tag{6.9}$$

Expression (6.8) is inserted into (6.9). To start with this iterative approximation of the ML estimators, at step $t = 1$ of the process, plausible values $N_{h,R}^{(1)}$ are used as estimates of N_h's. From this starting point, the algorithm generates estimates $N_{h,R}^{(t)}$ that converge to the ML estimators of $N_1, \ldots, N_H$ for $t \to \infty$.

Even when a moment estimator $N_{h,R}$ lies close to the boundaries of the natural parameter space, approximate confidence intervals based on the normal assumption might not be valid because the sampling distribution of such an estimator is likely non-symmetric. Alternatively, the finite population bootstrap procedure based on the generation of a bootstrap population (see Sect. 5) may serve as an instrument to calculate confidence intervals that include not only the sampling error but also the additional inaccuracy caused by the response randomization. For this purpose, the responses z_k can be used to generate the pseudo-population U_{boot}^* needed in this type of bootstrap process (see Fig. 5.1). In creating U_{boot}^*, value z_1 is replicated $\frac{1}{\pi_1}$ times, z_2 is replicated $\frac{1}{\pi_2}$ times, and so on. This gives a total of $N_{boot}^* = \sum_{sR} \frac{1}{\pi_k}$ replications that make up U_{boot}^*. Therein, with the y_k^i-values calculated by (6.1), the estimates $N_{h,R}$ are calculated according to (6.3). For the estimation of the sampling distributions of the $N_{h,R}$-values, B bootstrap samples are drawn from U_{boot}^* according to probability sampling method P with adapted first-order inclusion probabilities π_k^*. In each of these samples, variable z is observed. These observations are used to calculate B times either the moment or the ML estimators of $N_1, \ldots, N_H$. With the percentile method (Efron 1981, p. 317ff), the $(\alpha/2)$- and $(1 - \alpha/2)$-quantiles of the empirical distribution of the B estimators for each category size N_h are used to construct the respective confidence intervals at level $1 - \alpha$ (cf., for instance, van den Hout and van der Heijden 2002, p. 278).

6.2.3 Combining Direct and Randomized Responses

In practice, some of the respondents may be willing to divulge their true values of the sensitive variable y. Of course, the interviewees should not be asked "Do you want to disclose your true status or mask it?" In particular, for variables which are not variables that are sensitive as a whole like sexual behavior or income, but of which only the membership of a part of the possible categories is considered to be sensitive, to choose the option to scramble the response might be interpreted as admission to belong to one of these sensitive categories. This could apply, for instance, for variables such as drug usage or abortion. Therefore, the respondents would not feel well protected by offering this choice explicitly (for the issue of privacy protection, see Sect. 6.2.4). As a consequence, their willingness to participate in the survey would not increase.

But, it may often happen that, for instance, a telephone interviewer, while explaining the randomized response questioning design, recognizes that the interviewee is spontaneously willing to furnish the sensitive information directly without the randomization device (for example, by saying "Stop the explanations—I have no problem answering the question directly!"). Such an offer from a respondent to deliver the true y_k-value must not be ignored. Instead, to increase the efficiency of the estimators for the category sizes N_h $(h = 1, \ldots, H)$, the offer should be incorporated in the estimation procedure in the following way: Let population U theoretically be divided into (1) a subpopulation U_D of N_D elements willing to answer the direct question although an RR questioning design is offered to them, and (2) a disjoint group U_R, whose N_R members will use the offered randomization procedure $(U_D \cap U_R = \emptyset, U_D \cup U_R = U,$ and $N = N_D + N_R)$. Assuming full cooperation, the without-replacement probability sample s will then be divided into a group $s_D = s \cap U_D$ with n_D units delivering their true value y_k and a group $s_R = s \cap U_R$ with n_R elements with "missing values" $(s_R = s_m)$. These units will deliver a randomized response z_k. The true values y_k must be flagged in the data of the "mixed imputed sample" $s_I = s_D \cup s_R$ $(s_D \cap s_R = \emptyset, n = n_D + n_R)$.

This modification of a respondent's possible behavior affects the estimation procedure described in Sect. 6.2.1 in the following way.

Theorem 2a *For a probability sampling method P with arbitrary first-order sample inclusion probabilities π_k, a mixture (M) of direct answers y_k and randomized responses z_k collected using the RR model R,*

$$N_{h,M} = \sum_s{}' y'_{kh} \cdot \frac{1}{\pi_k} \qquad (6.10)$$

with

$$y'_{kh} = \begin{cases} y_{kh} & \text{if } k \in U_D, \\ y^i_{kh} & \text{otherwise} \end{cases}$$

$(h = 1, \ldots, H)$ is unbiased for the true population size N_h of category h of variable y in population U $(h = 1, \ldots, H)$ for any probability sampling technique P.

Proof Because $E_R(y^i_{kh}) = y_{kh}$, the expectation of (6.10) over P and R is equal to N_h.

For $s_D = \emptyset$ and $s = s_R$, $N_{h,M}$ reduces to $N_{h,R}$ (6.2), whereas for $s_R = \emptyset$ and $s = s_D$, the estimator $N_{h,M}$ reduces to the HT estimator $N_{h,\text{HT}} = \sum_s y_{kh} \cdot \frac{1}{\pi_k}$ of parameter N_h $(h = 1, \ldots, H)$.

The calculation of estimators $N_{h,M}$ of parameters N_h can be represented by the picture of a pseudo-population U_M^* of size $N_M^* = \sum_s \frac{1}{\pi_k}$ generated as a set-valued estimator of U with respect to these parameters $(h = 1, \ldots, H)$. In U_M^*, for each

Fig. 6.2 Generating a
pseudo-population with the
mixed questioning design M

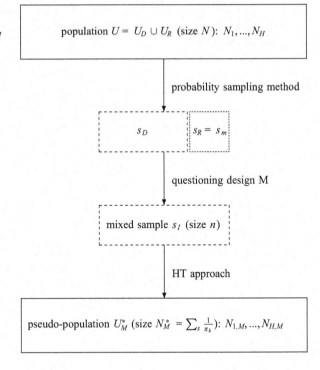

sampling element $k \in s_D$, a number of $\frac{1}{\pi_k}$ elements with variable values y_{kh} build a
direct answering part U_D^* of U_M^*, and for each $k \in s_R$, a number of $\frac{1}{\pi_k}$ elements with
variable values y_{kh}^i are generated, building the other part U_R^* of U_M^*. All replicated
variable values belong to the replication variable y_h^*. Summing the variable values
y_{kh}^* over the N_M^* units of $U_M^* = U_D^* \cup U_R^*$ ($U_D^* \cap U_R^* = \emptyset$) results in the estimator
$N_{h,M}$ from Eq. (6.10), which can also be explicated as

$$N_{h,M} = \sum_{U_M^*} y_{kh}^* \tag{6.11}$$

(see Fig. 6.2).

Theorem 2b *The theoretical variance of the estimator $N_{h,M}$ ($h = 1, \ldots, H$) is
given by*

$$V(N_{h,M}) = \sum\sum_U \Delta_{kl} \cdot \frac{y_{kh}}{\pi_k} \cdot \frac{y_{lh}}{\pi_l} + a_h \cdot \sum_{U_R} \frac{1}{\pi_k} + (b_h - 1) \cdot \sum_{U_R} y_{kh} \cdot \frac{1}{\pi_k}, \tag{6.12}$$

with a_h and b_h as defined in Theorem 1b.

Proof The theoretical variance of estimator (6.10) can be decomposed into

$$V(N_{h,M}) = V_P(E_R(N_{h,M}|s)) + E_P(V_R(N_{h,M}|s)).$$

For the first summand of the equation,

$$V_P(E_R(N_{h,M}|s)) = V_P\left(\sum_s y_{kh} \cdot \frac{1}{\pi_k}\right)$$

applies. Further,

$$E_P(V_R(N_{h,M}|s))$$

$$= E_P\left(V_R\left(\sum_{SD} y_{kh} \cdot \frac{1}{\pi_k}\bigg|s + \sum_{SR} y_{kh}^i \cdot \frac{1}{\pi_k}\bigg|s\right)\right)$$

$$= E_P\left(V_R\left(\sum_{UD} y_{kh} \cdot \frac{1}{\pi_k} \cdot I_k\bigg|s + \sum_{UR} y_{kh}^i \cdot \frac{1}{\pi_k} \cdot I_k\bigg|s\right)\right)$$

$$= E_P\left(\sum_{UR} \frac{1}{\pi_k^2} \cdot I_k^2 \cdot V_R(y_{kh}^i)\right).$$

With the development of $V_R(y_{kh}^i)$ presented in the proof of Theorem 1b,

$$E_P\left(\sum_{UR} \frac{1}{\pi_k^2} \cdot I_k^2 \cdot V_R(y_{kh}^i)\right) = a_h \cdot \sum_{UR} \frac{1}{\pi_k} + (b_h - 1) \cdot \sum_{UR} y_{kh} \cdot \frac{1}{\pi_k}$$

applies, which completes the proof of Theorem 2b.

The sum of the second and third summand of (6.12) is the price to be paid in terms of accuracy for the higher protection of the respondents' privacy when the "mixed strategy" M is applied. With $U_R = \emptyset$, (6.12) reduces to the variance of the HT estimator of the total N_h. For $U_D = \emptyset$, (6.12) reduces to (6.4).

Theorem 2c *The variance (6.12) can be estimated unbiasedly by*

$$\hat{V}(N_{h,M}) = \sum\sum_s \frac{\Delta_{kl}}{\pi_{kl}} \cdot \frac{y_{kh}'}{\pi_k} \cdot \frac{y_{lh}'}{\pi_l} + a_h \cdot \sum_{SR} \frac{1}{\pi_k} + (b_h - 1) \cdot \sum_{SR} y_{kh}^i \cdot \frac{1}{\pi_k}. \qquad (6.13)$$

Proof The expectation of the first summand of (6.13) over R is given by

$$E_R\left(\sum\sum_s \frac{\Delta_{kl}}{\pi_{kl}} \cdot \frac{y_{kh}'}{\pi_k} \cdot \frac{y_{lh}'}{\pi_l}\right)$$

$$= \sum\sum_U \frac{\Delta_{kl}}{\pi_{kl}} \cdot \frac{E_R(y_{kh}' \cdot y_{lh}')}{\pi_k \cdot \pi_l} \cdot I_k \cdot I_l$$

$$= \sum_U \frac{\Delta_{kk}}{\pi_k} \cdot \frac{E_R[(y_{kh}')^2]}{\pi_k^2} \cdot I_k^2 + \sum\sum_{U(k \neq l)} \frac{\Delta_{kl}}{\pi_{kl}} \cdot \frac{y_{kh} \cdot y_{lh}}{\pi_k \cdot \pi_l} \cdot I_k \cdot I_l.$$

For the expected value of $(y'_{kh})^2$,

$$E_R[(y'_{kh})^2] = \begin{cases} y_{kh} & \text{if } k \in U_D, \\ b_h \cdot y_{kh} + a_h & \text{otherwise} \end{cases}$$

applies. Hence, with $\Delta_{kk} = \pi_k \cdot (1 - \pi_k)$,

$$\sum_U (1 - \pi_k) \cdot \frac{E_R[(y'_{kh})^2]}{\pi_k^2} \cdot I_k^2 = \sum_{U_D} (1 - \pi_k) \cdot \frac{y_{kh}}{\pi_k^2} \cdot I_k^2$$

$$+ \sum_{U_R} (1 - \pi_k) \cdot \frac{a_h + b_h \cdot y_{kh}}{\pi_k^2} \cdot I_k^2$$

is derived. It follows that

$$E_P\left(E_R\left(\sum\sum_s \frac{\Delta_{kl}}{\pi_{kl}} \cdot \frac{y'_{kh}}{\pi_k} \cdot \frac{y'_{lh}}{\pi_l} \,\middle|\, s\right)\right)$$

$$= \sum_{U_D} (1 - \pi_k) \cdot \frac{y_{kh}}{\pi_k^2} \cdot E_P(I_k^2)$$

$$+ \sum_{U_R} (1 - \pi_k) \cdot \frac{a_h + b_h \cdot y_{kh}}{\pi_k^2} \cdot E_P(I_k^2)$$

$$+ \sum\sum_{U(k \neq l)} \frac{\Delta_{kl}}{\pi_{kl}} \cdot \frac{y_{kh}}{\pi_k} \cdot \frac{y_{lh}}{\pi_l} \cdot E_P(I_k \cdot I_l)$$

$$= \sum_{U_D} (1 - \pi_k) \cdot \frac{y_{kh}}{\pi_k} + \sum_{U_R} (1 - \pi_k) \cdot (a_h + b_h \cdot y_{kh}) \cdot \frac{1}{\pi_k}$$

$$+ \sum\sum_U \Delta_{kl} \cdot \frac{y_{kh}}{\pi_k} \cdot \frac{y_{lh}}{\pi_l} - \sum_U (1 - \pi_k) \cdot \frac{y_{kh}}{\pi_k}$$

$$= \sum\sum_U \Delta_{kl} \cdot \frac{y_{kh}}{\pi_k} \cdot \frac{y_{lh}}{\pi_l} + a_h \cdot \sum_{U_R} \frac{1 - \pi_k}{\pi_k}$$

$$+ (b_h - 1) \cdot \sum_{U_R} (1 - \pi_k) \cdot \frac{y_{kh}}{\pi_k}.$$

With

$$E_P\left(E_R\left(\sum_{s_R} (a_h + (b_h - 1) \cdot y'_{kh}) \cdot \frac{1}{\pi_k} \,\middle|\, s\right)\right)$$

$$= E_P\left(E_R\left(\sum_{U_R} (a_h + (b_h - 1) \cdot y'_{kh}) \cdot \frac{1}{\pi_k} \cdot I_k \,\middle|\, s\right)\right)$$

$$= \sum_{U_R} a_h + (b_h - 1) \cdot \sum_{U_R} y_{kh},$$

Theorem 2c is completely proven.

Theorem 2d *The theoretical covariance $C(N_{h,M}, N_{j,M})$ of "mixed" estimators $N_{h,M}$ and $N_{j,M}$ with categories $h \neq j$ is given by*

$$C(N_{h,M}, N_{j,M}) = -\frac{1}{v_h \cdot v_j} \cdot \sum_{U_R} (u_h \cdot v_j \cdot y_{kj} + u_j \cdot v_h \cdot y_{kh} + u_h \cdot u_j) \cdot \frac{1}{\pi_k}$$

$$+ \sum \sum_{U(k \neq l)} y_{kh} \cdot y_{lj} \cdot \frac{\pi_{kl}}{\pi_k \cdot \pi_l} - N_h \cdot N_j \qquad (6.14)$$

($h, j = 1, 2, \ldots, H$).

Proof The following applies

$$C(N_{h,M}, N_{j,M}) = E\left(\left(\sum_s y'_{kh} \cdot \frac{1}{\pi_k}\right) \cdot \left(\sum_s y'_{lj} \cdot \frac{1}{\pi_l}\right)\right) - N_h \cdot N_j$$

$$= E_P\left(E_R\left(\sum_s y'_{kh} \cdot y'_{kj} \cdot \frac{1}{\pi_k^2}\middle| s\right)\right)$$

$$+ E_P\left(E_R\left(\sum \sum_{s(k \neq l)} y'_{kh} \cdot y'_{lj} \cdot \frac{1}{\pi_k} \cdot \frac{1}{\pi_l}\middle| s\right)\right) - N_h \cdot N_j.$$

For

$$E_R(y'_{kh} \cdot y'_{kj}) = \begin{cases} 0 & \text{if } k \in U_D, \\ -\frac{u_h \cdot v_j \cdot y_{kj} + u_j \cdot v_h \cdot y_{kh} + u_h \cdot u_j}{v_h \cdot v_j} & \text{otherwise} \end{cases}$$

applies. For $k \neq l$, it follows that

$$E_R(y'_{kh} \cdot y'_{lj} | s) = E_R(y'_{kh} | s) \cdot E_R(y'_{lj} | s) = y_{kh} \cdot y_{lj}$$

applies. This yields

$$C(N_{h,M}, N_{j,M})$$

$$= E_P\left(-\frac{1}{v_h \cdot v_j} \cdot \sum_{U_R} (u_h \cdot v_j \cdot y_{kj} + u_j \cdot v_h \cdot y_{kh} + u_h \cdot u_j) \cdot \frac{1}{\pi_k^2} \cdot I_k\right)$$

$$+ E_P\left(\sum \sum_{U(k \neq l)} y_{kh} \cdot y_{lj} \cdot \frac{1}{\pi_k \cdot \pi_l} \cdot I_k \cdot I_l\right) - N_h \cdot N_j.$$

Because $E_P(I_k) = \pi_k$ and $E_P(I_k \cdot I_l) = \pi_{kl}$, the correctness of Theorem 2d is proven.

Theorem 2e *The covariance (6.14) of $N_{h,M}$ and $N_{j,M}$ is estimated unbiasedly by*

$$\hat{C}(N_{h,M}, N_{j,M}) = -\frac{1}{v_h \cdot v_j} \cdot \sum_{SR} (u_h \cdot v_j \cdot y^i_{kj} + u_j \cdot v_h \cdot y^i_{kh} + u_h \cdot u_j) \cdot \frac{1}{\pi_k^2}$$

$$+ \sum\sum_{s(k \neq l)} \frac{\Delta_{kl}}{\pi_{kl}} \cdot \frac{y'_{kh}}{\pi_k} \cdot \frac{y'_{lj}}{\pi_l} \tag{6.15}$$

$(h \neq j; h, j = 1, 2, \ldots, H)$.

Proof Based on the proof of Theorem 1e, the following expectations are needed:

$$E_P \left(E_R \left(\sum_{SR} y^i_{kh} \cdot \frac{1}{\pi_k^2} \middle| s \right) \right) = \sum_{U_R} y_{kh} \cdot \frac{1}{\pi_k}$$

and

$$E_P \left(E_R \left(\sum_{SR} \frac{1}{\pi_k^2} \middle| s \right) \right) = \sum_{U_R} \frac{1}{\pi_k}.$$

Hence, the unbiasedness of the first summand of (6.15) for the first summand of (6.14) is proven. Then,

$$E_P \left(E_R \left(\sum\sum_{s(k \neq l)} y'_{kh} \cdot y'_{lj} \cdot \frac{1}{\pi_k \cdot \pi_l} \middle| s \right) \right)$$

$$= E_P \left(\sum\sum_{s(k \neq l)} y_{kh} \cdot y_{lj} \cdot \frac{1}{\pi_k \cdot \pi_l} \cdot I_k \cdot I_l \middle| s \right)$$

$$= \sum\sum_{U(k \neq l)} y_{kh} \cdot y_{lj} \cdot \frac{\pi_{kl}}{\pi_k \cdot \pi_l}$$

applies. Moreover, in the same way as in the proof of Theorem 1e,

$$E_P \left(E_R \left(\sum\sum_{s(k \neq l)} y'_{kh} \cdot y'_{lj} \cdot \frac{1}{\pi_{kl}} \middle| s \right) \right)$$

$$= \sum\sum_{U(k \neq l)} y_{kh} \cdot y_{lj} = N_h \cdot N_j$$

applies. This completes the proof of the unbiasedness of $\hat{C}(N_{h,M}, N_{j,M})$ for $C(N_{h,M}, N_{j,M})$.

In summary, the randomization strategy R from Sect. 6.2.1 is a special case of the mixed strategy M allowing for true answers to the question on variable y, with $s_R = s$ and $U_R = U$. Using the variance and covariance estimators explicated in the present section, statistical hypotheses testing of differences in group sizes can be performed.

6.2.4 Accuracy and Privacy Protection

The efficiency of different strategies of RR depends strongly on the level of privacy protection that they offer. Therefore, in contrast to past claims in some publications, the efficiency of such strategies should be compared only at the same levels of privacy protection. Quatember (2009) showed the close connection between efficiency and privacy protection for RR techniques to estimate proportions. Naturally, this also applies for all special cases of the unified approach R (and M) discussed in the current chapter. Guerriero and Sandri (2007) described models with the same level of privacy protection as "equivalent" with respect to this protection (Guerriero and Sandri 2007, p. 2185). Let us define the following measures λ_h of the theoretical level of privacy protection according to the different categories h of the categorical variable y under study ($h = 1, \ldots, H$):

$$\lambda_h^{(R)} = \frac{P(z_k = h | y_k \neq h)}{P(z_k = h | y_k = h)}. \tag{6.16}$$

For the measure $\lambda_h^{(R)}$, the probability of answering "category h" ($z_k = h$), given that this is not the true category ($y_k \neq h$), is divided by the probability of this answer if this is the true category ($y_k = h$). Therefore, it refers to the privacy protection with respect to response $z_k = h$.

Assuming without loss of generality that $p_I > 0$, $p_I \geq p_{II,h}$ applies. Then, for the questioning design R, the "λ-measures" from (6.16) are given by

$$\lambda_h^{(R)} = \frac{p_{II,h} + p_{III,h} + p_{IV,h}}{p_I + p_{III,h} + p_{IV,h}} = \frac{u_h}{v_h + u_h} \tag{6.17}$$

with u_h, v_h as defined in Sect. 6.2.1 ($0 \leq \lambda_h^{(R)} \leq 1$, $h = 1, \ldots, H$). Hence, large $\lambda_h^{(R)}$-values indicate high theoretical privacy protection with respect to answer category h. The more $\lambda_h^{(R)}$ differs from one, the more information on the membership of group U_h is contained in the given answer z_k of element k and the less protected against disclosure the survey unit is with respect to variable y. Finally, for the direct questioning with p_I equal to one, $\lambda_h^{(R)} = 0$ apply for all categories $h = 1, \ldots, H$.

If the membership of a certain category h of variable y is objectively more sensitive than that of another category, its level $\lambda_h^{(R)}$ of privacy protection should be higher than that of the other categories with regard to the respondents' willingness to cooperate. In other words, the questioning design probabilities p_I, $p_{II,h}$ (with $y_k = i$ and $h \neq i$), $p_{III,h}$, and $p_{IV,h}$ according to the relative category sizes of auxiliary variable x can be chosen in accordance with the sensitivity levels of the different categories of variable y ($h = 1, \ldots, H$). This is relevant not only for the application of the questioning design R as randomized response technique but also for its application as a masking technique in the field of statistical disclosure control, where the probability mechanism R can be applied after the data collection to protect the privacy of respondents with regard to sensitive variables (see Sect. 7).

6.2.5 An Application to Opinion Polls

During the last decades, and not only in Austria (Europe), the results from opinion polls conducted shortly before a forthcoming election have increasingly differed from the actual outcome of the election. Particularly, the proportions of parties from the two margins of the political spectrum have constantly been underestimated. For this reason, representatives of such parties in Austria have accused opinion researchers of data manipulation. As a consequence, these politicians have demanded a prohibition of opinion polls shortly before elections. Actually, besides the increasingly late moment of voting decision, the growing sensitivity of the variable "voting behavior" and particularly of certain categories of this variable seem to be the main reason for this phenomenon.

For this application in the questioning design R, the design probabilities p_{II} and p_{IV} can be set to zero, whereas the others, p_{I} and $p_{\text{III},h}$ ($h = 1,\ldots,H$; $h \neq i$ with $y_k = i$), are freely selectable. This randomization setup is based on Liu and Chow (1976). In a telephone or face-to-face survey, such a questioning design can be implemented in the following manner (the procedure can also be easily adapted to Web or postal surveys): The interviewer tells the survey unit that because of the sensitivity of the subject, a questioning design will be applied to protect the respondents' privacy. Of course, the effect of this questioning design on data protection has to be explained to the respondents in a vivid way to produce the desired willingness to cooperate. Then, the respondent may be asked, for instance, to think of a person whose date of birth he or she knows but without delivering this information to the interviewer (for other random devices, see, for instance, Warner 1986, p. 441f). The ($H - 1$) parties in question and a non-voting category give altogether H possible answer categories. Then, if the date of birth is within a certain interval, such as from January to September, the respondent shall answer truthfully a question such as, "Imagine it's election day. Which party gets your vote?" However, if the date is, say, from the 1st to the 20th of October, the respondent shall simply answer "party 1." If the date lies within an interval, say, from the 21st of October to the 9th of November, the survey unit shall simply answer "party 2," and so on. In any case, these disjoint date groups have to cover all possible dates of birth from January to December. The chosen allocation of these dates to groups determines the design probabilities. For simplicity, the probabilities for certain dates may, for example, be approximated by a uniform distribution.

A mathematically more sophisticated randomizing device with no uniform distribution that requires no instrument such as a dice was suggested by Diekmann (2012) and makes use of the Newcomb–Benford distribution (cf. Newcomb 1881). For this purpose, the respondent may be asked to think of a person of whom he or she recalls the house number. Then, if the first digit of the house number is within a certain interval as from 1 to 4, the respondent shall answer truthfully on the interesting question from above. But, if it is, say, 5 or 6, the respondent shall answer "party 1." If the digit is 7, the survey unit shall answer "party 2," and so on. After all, the H groups of first digits have to cover all nine numbers. The probability of a certain first digit follows the Newcomb–Benford distribution. Therefore, the probability of being asked the direct question is 0.699 (cf. Newcomb 1881, p. 40). The probability of 5 or 6 is 0.146, of 7 is 0.058, of 8 is 0.051, and of 9 it is the probability of 0.046. If other probabilities are needed for the questioning design or the number H of answer categories is larger, the grouping of the first digits should be done in another way, or the first two digits can be used to produce more possible groups (cf. Newcomb 1881, p. 40). Diekmann (2012) emphasizes that for this random device there is a discrepancy between the probabilities as perceived by the respondents and the actual probabilities. For instance, the probability of picking a house number with first digit from 1 to 4 is believed to be smaller than it actually is. This "illusion" (Diekmann 2012, p. 330) has a positive effect on the perceived privacy protection with regard to this RR questioning design (for a discussion on the different aspects of theoretical and perceived privacy protection, see Chaudhuri and Christofides 2013, Chap. 7).

If during the design explanations, a respondent reveals without being asked to do so that he or she is willing to answer the sensitive question directly (as mentioned, for instance, in Sect. 6.2.3), this answer must be flagged. Then the estimator (6.10) has to be applied.

If the direct questioning on the sensitive variable leads to non-ignorable non-response and untruthful answering, as expected in opinion polls, a considerably biased estimator is the consequence. For such cases, the higher complexity of the RR questioning design R will surely pay off under the assumption of cooperation. The accuracy of the estimators increases although their variances exceed the (then only) theoretical variances of the direct questioning.

6.3 The Estimation of a Total by the Randomized Response Technique

6.3.1 A Generalized Questioning Design for Quantitative Variables

The idea of reducing nonresponse and untruthful answering by applying an alternative questioning design that protects the respondent's privacy need not to be restricted to categorical variables. The first effort to apply such a method also to quantitative variables was undertaken by Greenberg et al. (1971). It was a direct development from the theoretical framework given by Greenberg et al. (1969) for the estimation of proportions applying the model by Horvitz et al. (1967). To estimate the mean $\bar{y}$ of a quantitative variable y (such as the number of abortions per unit in a given population of women), it was suggested to ask a respondent k either with probability p_I for the true value y_k or with the remaining probability p_{II} for the true value x_k of a completely innocuous quantitative variable x unrelated to y. Variable x should have a similar range of possible values as y. This means that for the technique discussed by Greenberg et al. (1971) the answer $z_k^{(G)}$ of a respondent k is

$$z_k^{(G)} = \begin{cases} y_k & \text{with probability } p_{\mathrm{I}}, \\ x_k & \text{with probability } p_{\mathrm{II}} = 1 - p_{\mathrm{I}} \end{cases}$$

$(0 \le p_{\mathrm{I}} \le 1)$. In practice, the design probabilities p_{I} and p_{II} can be implemented by rotating a spinner, drawing cards, throwing the dice, building disjoint date groups (see Sect. 6.2.5) or using a computer program as randomization instrument (cf. Lensvelt-Mulders et al. 2005). Obviously, it is more efficient for the estimation of $\bar{y}$ to know the population distribution of x with expectation μ_x and variance σ_x^2. If these parameters are not known, a two-sample procedure can be applied, in which the statistical characteristics of x have to be estimated as well (cf. Greenberg et al. 1971, p. 245).

After this first application of an RR questioning design to quantitative variables, other techniques have been developed for the same purpose (cf. for a review, for instance, Diana and Perri 2011, p. 635ff). Again, after the presentation of basic ideas, other developments consist of chaining and combining these techniques (recent examples are Bar-Lev et al. 2004; Gjestvang and Singh 2007). In the following, a generalized framework for the estimation of the total t of a (sensitive) variable y in general probability sampling is presented. It includes, as special cases, the most important techniques already published. Furthermore, this unified approach at the same time encompasses all the other combinations not yet published. The execution of the statistical properties of this framework for general probability sampling, being of the greatest importance for the practical application of the procedure, is based on unbiased "imputations" y_k^i for the true values y_k ($k \in s$) and the HT approach of generating a pseudo-population.

The proposed unification Q of different RR questioning designs for the estimation of the total t of a quantitative variable y is described in the following way: A drawn sampling unit k is asked

- with an overall probability p_{I} for his or her true value y_k of variable y (Instruction I);
- with probability p_{II} for his or her true value x_k of a nonsensitive auxiliary variable x unrelated to y (Instruction II);
- with probability p_{III} to answer with the result of $y_k + u_k$ where u is a random variable with known distribution, expected value μ_u, and variance σ_u^2 (Instruction III);
- with probability p_{IV} to answer with the result of $y_k \cdot v_k$ where v is a random variable with known distribution, expected value μ_v, and variance σ_v^2 (Instruction IV); or
- with probability p_{V} to answer with value w_k of random variable w with known distribution, expected value μ_w, and variance σ_w^2 (Instruction V).

This means that for the RR framework Q technique the answer z_k of a respondent k is

$$z_k = \begin{cases} y_k & \text{with probability } p_{\mathrm{I}}, \\ x_k & \text{with probability } p_{\mathrm{II}}, \\ y_k + u_k & \text{with probability } p_{\mathrm{III}}, \\ y_k \cdot v_k & \text{with probability } p_{\mathrm{IV}}, \\ w_k & \text{with probability } p_{\mathrm{V}} \end{cases}$$

($\sum p_i = 1$). As it was described for the questioning design R in Sect. 6.2.1, the probability p_{I} of being asked the direct question on variable y can be composed of different probabilities from $K \geq 1$ stages (cf., for instance, Ryu et al. 2005). This can be performed to increase the respondents' subjective perception of the privacy protection offered by the RR technique, which is considered as the "respondents' criterion for participation in an indirect questioning survey" (Chaudhuri and Christofides 2013, p. 151f). The procedure might start at stage 1, at which the question on y is asked with probability $p_{\mathrm{I}}^{(1)} < 1$. With the remaining probability, the sampling unit is directed to a second stage and so on. At the final Kth stage, where all the five instructions of Q are possible, the survey unit is asked the question on y with probability $p_{\mathrm{I}}^{(K)}$. This gives the overall probability

$$p_{\mathrm{I}} = p_{\mathrm{I}}^{(1)} + (1 - p_{\mathrm{I}}^{(1)}) \cdot p_{\mathrm{I}}^{(2)} + \cdots + (1 - p_{\mathrm{I}}^{(1)}) \cdot (1 - p_{\mathrm{I}}^{(2)}) \cdot \ldots \cdot (1 - p_{\mathrm{I}}^{(K-1)}) \cdot p_{\mathrm{I}}^{(K)}$$

of being asked the question on variable y under study in the multi-stage questioning design already mentioned in Sect. 6.2.1. The auxiliary variable x needed in Instruction II is the one adopted by Greenberg et al. (1971). The variables u and v are scrambling random variables to mask the true value y_k of a survey unit k. Lastly, w is a random variable predetermined by the agency with respect to the possible values of y. Instruction V differs from Instruction II in the random nature of the mechanism that leads to answers x_k and w_k, respectively. On the one hand, the value of variable x is fixed for a given sample unit k. Its random nature is solely the result of the selection of a probability sample. On the other hand, the value of variable w is random for each sample element. Note that the random variables u, v, and w should be independent, but could have the same distributions.

Again, to avoid any misunderstanding about the meaning of this presentation, this is not a proposal to use all of the instructions included in questioning design Q at the same time. But based on the unified framework Q, the theoretical properties of a total estimator can be derived under one theoretical roof for any combination of the five instructions of Q. In the following, some of these combinations, which to the author's best knowledge have already been published are mentioned as examples.

- For $p_{\mathrm{I}} = 1$, the direct questioning on the subject is included in the RR technique Q.

- For $p_{III} = 1$, the addition strategy proposed by Pollock and Bek (1976) is part of the generalization Q.
- For $p_{IV} = 1$, the multiplicative model proposed and theoretically explicated by Poole (1974), Pollock and Bek (1976), and Eichhorn and Hayre (1983) and its two-stage version by Ryu et al. (2005) is a special case of Q.

Moreover, the following ideas are also special cases of Q. For all of them $0 < p_I < 1$ applies and those design probabilities from p_{II} to p_V not mentioned explicitly are set to zero:

- For $p_{II} = 1 - p_I$, the strategy presented by Greenberg et al. (1971) with known distribution of x is a member of family Q.
- With $p_{IV} = 1 - p_I$, the strategy presented by Bar-Lev et al. (2004) belongs to our framework Q.
- With $p_{IV}, p_V > 0$, $p_{IV} + p_V = 1 - p_I$, and w being a "random variable" with a fixed value, the model of Gjestvang and Singh (2007) is also an incorporated member of Q.

To explicate the theoretical discussion of RR questioning design Q for general probability sampling with arbitrary first-order sample inclusion probabilities π_k ($k \in U$), the expected value of the response z_k of a certain survey unit k for given y results in

$$E_Q(z_k) = p_I \cdot y_k + p_{II} \cdot \mu_x + p_{III} \cdot (y_k + \mu_u) + p_{IV} \cdot y_k \cdot \mu_v + p_V \cdot \mu_w$$
$$= y_k \cdot \underbrace{(p_I + p_{III} + p_{IV} \cdot \mu_v)}_{\equiv b} + \underbrace{p_{II} \cdot \mu_x + p_{III} \cdot \mu_u + p_V \cdot \mu_w}_{\equiv a}.$$

Obviously, the term

$$y_k^i = \frac{z_k - a}{b} \tag{6.18}$$

($b \neq 0$) is unbiased for the true value y_k of unit k over the randomization Q.

The development of RR technique Q for a general without-replacement probability sampling technique P can make use of these substitutes y_k^i for the unknown y_k-values to generate a pseudo-population, say, U_Q^* as estimator of the original population U with respect to the estimation of t from an imputed sample s_I (see Sect. 3.3). Assuming that the higher privacy protection offered by questioning design Q promotes the respondents' motivation to cooperate, the following theorems apply:

Theorem 3a *For a probability sampling method P with arbitrary first-order sample inclusion probabilities π_k ($k \in U$) and questioning design Q, the total t of study variable y in population U is unbiasedly estimated by*

$$t_Q = \sum_s y_k^i \cdot \frac{1}{\pi_k}. \tag{6.19}$$

Fig. 6.3 Generating a
pseudo-population with the
questioning design Q

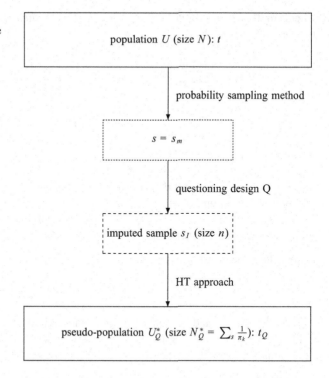

Proof It is easy to show the unbiasedness of estimator (6.19) because $E_Q(y_k^i)$
denoting the expectation of y_k^i over the randomization process Q equals y_k and the
HT estimator $\sum_s y_k \cdot \frac{1}{\pi_k}$ is unbiased over sampling method P.

Hence, the estimation of t by (6.19), which is a special case of the imputation
estimator t_I (3.4), can be interpreted as the generation of a pseudo-population U_Q^*.
For this purpose, the sampling units drawn by a probability sampling scheme P are
subject to the randomization procedure Q. This probability mechanism that masks
the true values of variable y creates a sample s with imputed values y_k^i, instead of
the true values y_k ($k \in s$). According to Eq. (6.19), the artificial population U_Q^* is
generated from s_Q by replicating each of the n sampled y_k^i-values exactly $\frac{1}{\pi_k}$ times
(see Fig. 6.3). The size N_Q^* of pseudo-population U_Q^* is equal to the sum $\sum_s \frac{1}{\pi_k}$.
Hence, N_Q^* depends on the probability sampling method P applied. The estimator t_Q
is the sum of all values y_k^{i*} ($k \in U_Q^*$) denoting the values y_k^i replicated from s_Q to
generate U_Q^*:

$$t_Q = \sum_{U_Q^*} y_k^{i*}. \tag{6.20}$$

Theorem 3b *For a probability sampling method P, the theoretical variance of t_Q is given by*

$$V(t_Q) = \sum\sum_U \Delta_{kl} \cdot \frac{y_k}{\pi_k} \cdot \frac{y_l}{\pi_l} + \frac{1}{b^2} \cdot \left[(c - b^2) \cdot \sum_U y_k^2 \cdot \frac{1}{\pi_k} \right.$$

$$\left. + 2 \cdot (d - a \cdot b) \cdot \sum_U y_k \cdot \frac{1}{\pi_k} + (e - a^2) \cdot \sum_U \frac{1}{\pi_k} \right] \qquad (6.21)$$

with $c \equiv p_I + p_{III} + p_{IV} \cdot (\sigma_v^2 + \mu_v^2)$, $d \equiv p_{III} \cdot \mu_u$, *and* $e \equiv p_{II} \cdot (\sigma_x^2 + \mu_x^2) + p_{III} \cdot (\sigma_u^2 + \mu_u^2) + p_V \cdot (\sigma_w^2 + \mu_w^2)$. *The first summand of variance (6.21) refers to the variance of the HT estimator for the total of variable y for a given probability sampling scheme P, when the question on variable y is asked directly. Then, the second summand in (6.21) can be seen as the price to be paid in terms of accuracy for the privacy protection offered by the questioning design Q to the respondents.*

Proof The variance of t_Q is given by

$$V(t_Q) = V_P(E_Q(t_Q|s)) + E_P(V_Q(t_Q|s)). \qquad (6.22)$$

Herein, E_P, V_P, E_Q, and V_Q denote the expectations and variances over the two random processes involved, the probability sampling design P, and the randomization of responses Q. The first of the two summands on the right-hand side of (6.22) yields

$$V_P(E_Q(t_Q|s)) = V_P\left(\sum_s y_k \cdot \frac{1}{\pi_k} \right) = \sum\sum_U \Delta_{kl} \cdot \frac{y_k}{\pi_k} \cdot \frac{y_l}{\pi_l}. \qquad (6.23)$$

Let I_k again indicate the sample inclusion of survey unit k ($k = 1, \ldots, N$). Because the covariance C_R of the substitutes y_k^i and y_l^i ($k \neq l$) over the randomization Q is zero, the variance of t_Q over the randomization Q conditioned on sample s is given by

$$V_Q(t_Q|s) = V_Q\left(\sum_U y_k^i \cdot \frac{1}{\pi_k} \cdot I_k \,\middle|\, s \right) = \sum_U I_k^2 \cdot \frac{1}{\pi_k^2} \cdot V_Q(y_k^i).$$

Hence, with $E_P(I_k^2) = \pi_k$, for the second summand on the right-hand side of (6.22),

$$E_P(V_Q(t_Q|s)) = \sum_U V_Q(y_k^i) \cdot \frac{1}{\pi_k}$$

applies with

$$V_Q(y_k^i) = \frac{1}{b^2} \cdot V_Q(z_k).$$

Moreover, $V_Q(z_k) = E_Q(z_k^2) - E_Q^2(z_k)$. The first expected value yields

$$
\begin{aligned}
E_Q(z_k^2) &= p_I \cdot y_k^2 + p_{II} \cdot (\sigma_x^2 + \mu_x^2) + p_{III} \cdot (y_k^2 + 2 \cdot y_k \cdot \mu_u + \sigma_u^2 + \mu_u^2) \\
&\quad + p_{IV} \cdot y_k^2 \cdot (\sigma_v^2 + \mu_v^2) + p_V \cdot (\sigma_w^2 + \mu_w^2) \\
&= y_k^2 \cdot \underbrace{(p_I + p_{III} + p_{IV} \cdot (\sigma_v^2 + \mu_v^2))}_{\equiv c} + 2 \cdot \underbrace{p_{III} \cdot \mu_u}_{\equiv d} \cdot y_k \\
&\quad + \underbrace{p_{II} \cdot (\sigma_x^2 + \mu_x^2) + p_{III} \cdot (\sigma_u^2 + \mu_u^2) + p_V \cdot (\sigma_w^2 + \mu_w^2)}_{\equiv e}
\end{aligned}
$$

With $E_Q^2(z_k) = (y_k \cdot b + a)^2$, variance $V_Q(z_k)$ is given by

$$
V_Q(z_k) = y_k^2 \cdot (c - b^2) + 2 \cdot y_k \cdot (d - a \cdot b) + e - a^2.
$$

Hence, the expectation $E_P(V_Q(t_Q|s))$ on the right-hand side of Eq. (6.22) results in

$$
\begin{aligned}
E_P(V_Q(t_Q|s)) = \frac{1}{b^2} \cdot \Big[(c - b^2) \cdot \sum_U y_k^2 \cdot \frac{1}{\pi_k} \\
+ 2 \cdot (d - a \cdot b) \cdot \sum_U y_k \cdot \frac{1}{\pi_k} + (e - a^2) \cdot \sum_U \frac{1}{\pi_k} \Big].
\end{aligned}
$$

Summing this up with (6.23) proves Theorem 3b.

From the first derivative after μ_w, for all other selectable options given, the variance minimizing function of expected value μ_w, which can be fixed by the agency, results in

$$
\mu_w = \frac{b \cdot \sum_U y_k \cdot \frac{1}{\pi_k} + (p_{II} \cdot \mu_x + d) \cdot \sum_U \frac{1}{\pi_k}}{(1 - p_V) \cdot \sum_U \frac{1}{\pi_k}}. \tag{6.24}
$$

In (6.24), observations from past surveys may help to estimate the sum $\sum_U y_k \cdot \frac{1}{\pi_k}$ needed in the enumerator. It can be estimated by $\sum_s y_k \cdot \frac{1}{\pi_k^2}$ from a direct questioning design and by $\sum_s y_k^i \cdot \frac{1}{\pi_k^2}$ from an RR questioning design Q, respectively. Applying the probability mechanism of randomization model Q as a method of statistical disclosure control after the data collection (see the subsequent chapter), with (6.24), the agency can minimize the variance of a total estimator calculated from the masked data, when Instruction V is used in the masking process.

Theorem 3c *An unbiased estimator of the theoretical variance $V(t_Q)$ according to (6.21) is given by*

$$
\begin{aligned}
\hat{V}(t_Q) = \sum\sum_s \frac{\Delta_{kl}}{\pi_{kl}} \cdot \frac{y_k^i}{\pi_k} \cdot \frac{y_l^i}{\pi_l} + \frac{1}{c} \cdot \Big[(c - b^2) \cdot \sum_s (y_k^i)^2 \cdot \frac{1}{\pi_k} \\
+ 2 \cdot (d - a \cdot b) \cdot \sum_s y_k^i \cdot \frac{1}{\pi_k} + (e - a^2) \cdot \sum_s \frac{1}{\pi_k} \Big].
\end{aligned} \tag{6.25}
$$

Proof The expectation of the first summand over both processes P and Q yields

$$E_Q \left(\sum \sum_s \frac{\Delta_{kl}}{\pi_{kl}} \cdot \frac{y_k^i}{\pi_k} \cdot \frac{y_l^i}{\pi_l} \right) = \sum \sum_U \frac{\Delta_{kl}}{\pi_{kl}} \cdot \frac{E_Q(y_k^i \cdot y_l^i)}{\pi_k \cdot \pi_l} \cdot I_k \cdot I_l$$

$$= \sum \sum_{U(k \neq l)} \frac{\Delta_{kl}}{\pi_{kl}} \cdot \frac{y_k}{\pi_k} \cdot \frac{y_l}{\pi_l} \cdot I_k \cdot I_l$$

$$+ \sum_U \frac{\Delta_{kk}}{\pi_k} \cdot \frac{E_Q[(y_k^i)^2]}{\pi_k^2} \cdot I_k^2.$$

Therein,

$$E_Q[(y_k^i)^2] = \frac{1}{b^2} \cdot E_Q \left(z_k^2 - 2 \cdot a \cdot z_k + a^2 \right)$$

$$= \frac{1}{b^2} \cdot \left(E_Q(z_k^2) - 2 \cdot a \cdot (y_k \cdot b + a) + a^2 \right)$$

applies. With the notations from the proof of Theorem 3b, the expected value of the squared responses z_k^2 over the RR strategy Q is given by

$$E_Q(z_k^2) = y_k^2 \cdot c + 2 \cdot y_k \cdot d + e.$$

Thus,

$$E_Q[(y_k^i)^2] = \frac{1}{b^2} \cdot \left(y_k^2 \cdot c + 2 \cdot y_k \cdot (d - a \cdot b) + e - a^2 \right)$$

applies. Inserting this term in the development of $E_Q \left(\sum \sum_s \frac{\Delta_{kl}}{\pi_{kl}} \cdot \frac{y_k^i}{\pi_k} \cdot \frac{y_l^i}{\pi_l} \right)$ provides

$$E_P \left(E_Q \left(\sum \sum_s \frac{\Delta_{kl}}{\pi_{kl}} \cdot \frac{y_k^i}{\pi_k} \cdot \frac{y_l^i}{\pi_l} \middle| s \right) \right)$$

$$= \sum \sum_{U(k \neq l)} \frac{\Delta_{kl}}{\pi_{kl}} \cdot \frac{y_k}{\pi_k} \cdot \frac{y_l}{\pi_l} \cdot E_P(I_k \cdot I_l)$$

$$+ \frac{1}{b^2} \cdot \sum_U \frac{\Delta_{kk}}{\pi_k} \cdot \frac{\left(y_k^2 \cdot c + 2 \cdot y_k \cdot (d - a \cdot b) + e - a^2 \right)}{\pi_k^2} \cdot E_P(I_k^2)$$

$$= \sum \sum_U \Delta_{kl} \cdot \frac{y_k}{\pi_k} \cdot \frac{y_l}{\pi_l} - \sum_U \Delta_{kk} \cdot \frac{y_k^2}{\pi_k^2} + \frac{c}{b^2} \cdot \sum_U \Delta_{kk} \cdot y_k^2 \cdot \frac{1}{\pi_k^2}$$

$$+ 2 \cdot \frac{d - a \cdot b}{b^2} \cdot \sum_U \Delta_{kk} \cdot y_k \cdot \frac{1}{\pi_k^2} + \frac{e - a^2}{b^2} \cdot \sum_U \frac{\Delta_{kk}}{\pi_k^2}$$

$$= \sum \sum_U \Delta_{kl} \cdot \frac{y_k}{\pi_k} \cdot \frac{y_l}{\pi_l} + \frac{1}{b^2} \left[(c - b^2) \cdot \sum_U (1 - \pi_k) \cdot y_k^2 \cdot \frac{1}{\pi_k} \right.$$

$$\left. + 2 \cdot (d - a \cdot b) \cdot \sum_U (1 - \pi_k) \cdot y_k \cdot \frac{1}{\pi_k} + (e - a^2) \cdot \sum_U \frac{1 - \pi_k}{\pi_k} \right].$$

Hence, to get an unbiased estimator of the variance (6.21), it is necessary to add terms, of which the expectations over P and Q yield $\frac{1}{b^2} \cdot (c - b^2) \cdot \sum_U y_k^2$, $\frac{1}{b^2} \cdot 2 \cdot (d - a \cdot b) \cdot \sum_U y_k$, and $\frac{1}{b^2} \cdot (e - a^2) \cdot N$. With respect to the first of these three terms, with $E_Q[(y_k^i)^2] = \frac{1}{b^2} \cdot (y_k^2 \cdot c + 2 \cdot y_k \cdot (d - a \cdot b) + e - a^2)$, the following applies:

$$E_P \left(E_Q \left(\frac{1}{c} \cdot \sum_s ((y_k^i)^2 \cdot b^2 - 2 \cdot y_k^i \cdot (d - a \cdot b) - e + a^2) \cdot \frac{1}{\pi_k} \middle| s \right) \right)$$

$$= E_P \left(\sum_s y_k^2 \cdot \frac{1}{\pi_k} + 2 \cdot \frac{d - a \cdot b}{c} \cdot \sum_s y_k \cdot \frac{1}{\pi_k} + \frac{e - a^2}{c} \cdot \sum_s \frac{1}{\pi_k} \right.$$

$$\left. - 2 \cdot \frac{d - a \cdot b}{c} \cdot \sum_s y_k \cdot \frac{1}{\pi_k} - \frac{e - a^2}{c} \cdot \sum_s \frac{1}{\pi_k} \right)$$

$$= E_P \left(\sum_s y_k^2 \cdot \frac{1}{\pi_k} \right) = \sum_U y_k^2.$$

Furthermore,

$$E_P \left(E_Q \left(\sum_s y_k^i \cdot \frac{1}{\pi_k} \middle| s \right) \right) = \sum_U y_k$$

applies and the expected value of $\sum_s \frac{1}{\pi_k}$ is equal to N. Therefore,

$$\hat{V}(t_Q) = \sum_s \sum_s \frac{\Delta_{kl}}{\pi_{kl}} \cdot \frac{y_k^i}{\pi_k} \cdot \frac{y_l^i}{\pi_l} + \frac{1}{b^2} \cdot (c - b^2) \cdot \frac{1}{c} \cdot \sum_s b^2 \cdot (y_k^i)^2 \cdot \frac{1}{\pi_k}$$

$$- \frac{1}{b^2} \cdot (c - b^2) \cdot \frac{1}{c} \cdot 2 \cdot (d - a \cdot b) \cdot \sum_s y_k^i \cdot \frac{1}{\pi_k}$$

$$- \frac{1}{b^2} \cdot (c - b^2) \cdot \frac{1}{c} \cdot (e - a^2) \cdot \sum_s \frac{1}{\pi_k}$$

$$+ \frac{1}{b^2} \cdot 2 \cdot (d - a \cdot b) \cdot \sum_s y_k^i \cdot \frac{1}{\pi_k} + \frac{1}{b^2} \cdot (e - a^2) \cdot \sum_s \frac{1}{\pi_k}$$

is unbiased for $V(t_Q)$. Summing up the components of this equation yields (6.25).

6.3.2 Combining Direct and Randomized Responses

As mentioned in Sect. 6.2.3, in the practice of statistical surveys conducted, for example, by market or opinion research agencies, it may happen that some of the respondents are willing to deliver the true value of the sensitive variable without being asked to do so by saying something like "I have no problem answering the question directly." To not take up the offer and continue the RR strategy Q

undeterred to mask the true answer would make no sense in such cases. To take the offer and use the knowledge that the answer z_k of a certain survey unit k is y_k has to be incorporated in the theoretical properties of the estimator for t because the variance of the estimator will surely decrease when such responding behavior happens.

For this purpose, let population U theoretically be divided into (1) a subpopulation U_D of N_D elements willing to answer the direct question although an RR questioning design is offered to them, and (2) a disjoint group U_Q of size N_Q, whose members will use the offered randomization procedure Q ($U_D \cap U_Q = \emptyset$, $U = U_D \cup U_Q$, and $N = N_D + N_Q$). The partitioning of the without-replacement probability sample s, as described in Sect. 6.2.3, into a group s_D of survey units answering the direct question and a second group s_Q answering the RR technique Q is incorporated in the questioning design Q in the following way. Note that the values y_k of the subgroup s_D of the sample s have to be identifiable within the data set.

Theorem 4a *For a without-replacement probability sampling method P and a mixture (L) of direct answers y_k and randomized responses z_k collected using the RR model Q,*

$$t_L = \sum_s y'_k \cdot \frac{1}{\pi_k} \tag{6.26}$$

with

$$y'_k = \begin{cases} y_k & \text{if } k \in U_D, \\ y^i_k & \text{otherwise} \end{cases}$$

is unbiased for the total t of variable y in population U for any probability sampling scheme P.

Proof Because $E_Q(y^i_k) = y_k$, the expected value of (6.26) over P and Q results in

$$E_P(E_Q(t_L)) = E_P \left(\sum_s y_k \cdot \frac{1}{\pi_k} \right) = t.$$

For $s_D = \emptyset$ and $s = s_Q$, t_L reduces to t_Q, whereas for $s_Q = \emptyset$ and $s = s_D$, estimator t_L reduces to the HT estimator $t_{HT} = \sum_s y_k \cdot \frac{1}{\pi_k}$ of parameter t. For the estimation process represented by formula (6.26), a pseudo-population U^*_L of size $N^*_L = \sum_s \frac{1}{\pi_k}$ is generated according to the HT approach. The y^*-values of the N^*_L elements of U^*_L are created by a replication of the sample $s = s_D + s_Q$ and, therefore, correspond to either the true value y_k or the imputed value y^i_k (see Fig. 6.4). Hence, t_L can be written by

$$t_L = \sum_{U^*_L} y^*_k. \tag{6.27}$$

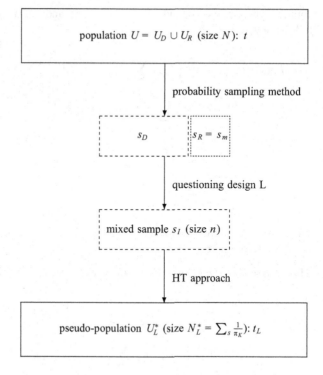

Fig. 6.4 Generating a pseudo-population with the mixed questioning design L

Theorem 4b *The theoretical variance of t_L for a sample conducted by a probability sampling scheme P is given by*

$$V(t_L) = \sum\sum_U \Delta_{kl} \cdot \frac{y_k}{\pi_k} \cdot \frac{y_l}{\pi_l} + \frac{1}{b^2} \cdot \left[(c - b^2) \cdot \sum_{U_Q} y_k^2 \cdot \frac{1}{\pi_k} \right.$$

$$\left. + 2 \cdot (d - a \cdot b) \cdot \sum_{U_Q} y_k \cdot \frac{1}{\pi_k} + (e - a^2) \cdot \sum_{U_Q} \frac{1}{\pi_k} \right] \quad (6.28)$$

with c, d, and e as defined in Theorem 3b.

Proof The variance of t_L is given by

$$V(t_L) = V_P(E_Q(t_L|s)) + E_P(V_Q(t_L|s)).$$

For

$$V_P(E_Q(t_L|s)) = V_P\left(\sum_s y_k \cdot \frac{1}{\pi_k} \right) = \sum\sum_U \Delta_{kl} \cdot \frac{y_k}{\pi_k} \cdot \frac{y_l}{\pi_l}$$

applies, whereas the second of the two summands of $V(t_L)$ is developed in the following way:

$$E_P(V_Q(t_L|s)) = E_P\left(V_Q\left(\sum_U y'_k \cdot \frac{1}{\pi_k} \cdot I_k \middle| s\right)\right)$$

$$= E_P\left(\sum_{U_Q} I_k^2 \cdot \frac{1}{\pi_k^2} \cdot V_Q(y'_k)\right)$$

$$= \frac{1}{b^2} \cdot \left[(c - b^2) \cdot \sum_{U_Q} y_k^2 \cdot \frac{1}{\pi_k}\right.$$

$$+ 2 \cdot (d - a \cdot b) \cdot \sum_{U_Q} y_k \cdot \frac{1}{\pi_k} + (e - a^2) \cdot \sum_{U_Q} \frac{1}{\pi_k}\right].$$

This results from the derivations for the proof of Theorem 3b and completes the proof of Theorem 4b.

Theorem 4c *An unbiased estimator of the theoretical variance $V(t_L)$ is given by*

$$\hat{V}(t_L) = \sum\sum_s \frac{\Delta_{kl}}{\pi_{kl}} \cdot \frac{y'_k}{\pi_k} \cdot \frac{y'_l}{\pi_l} + \frac{1}{c} \cdot \left[(c - b^2) \cdot \sum_{sQ} (y'_k)^2 \cdot \frac{1}{\pi_k}\right.$$

$$+ 2 \cdot (d - a \cdot b) \cdot \sum_{sQ} y_k^i \cdot \frac{1}{\pi_k} + (e - a^2) \cdot \sum_{sQ} \frac{1}{\pi_k}\right]. \qquad (6.29)$$

Proof The expectation of the first summand over the randomization strategy Q yields

$$E_Q\left(\sum\sum_s \frac{\Delta_{kl}}{\pi_{kl}} \cdot \frac{y'_k}{\pi_k} \cdot \frac{y'_l}{\pi_l}\right) = \sum\sum_U \frac{\Delta_{kl}}{\pi_{kl}} \cdot \frac{E_Q(y'_k \cdot y'_l)}{\pi_k \cdot \pi_l} \cdot I_k \cdot I_l$$

$$= \sum\sum_{U(k \neq l)} \frac{\Delta_{kl}}{\pi_{kl}} \cdot \frac{y_k}{\pi_k} \cdot \frac{y_l}{\pi_l} \cdot I_k \cdot I_l$$

$$+ \sum_U \frac{\Delta_{kk}}{\pi_k} \cdot \frac{E_Q[(y'_k)^2]}{\pi_k^2} \cdot I_k^2.$$

Moreover, the expected value of $(y'_{kh})^2$ over Q is given by

$$E_Q[(y'_k)^2] = \begin{cases} y_k^2 & \text{if } k \in U_D, \\ \frac{1}{b^2} \cdot (c \cdot y_k^2 + 2 \cdot (d - a \cdot b) \cdot y_k + e - a^2) & \text{otherwise.} \end{cases}$$

Hence,

$$\sum_U (1 - \pi_k) \cdot \frac{E_Q[(y_k')^2]}{\pi_k^2} \cdot I_k^2 = \sum_{U_D} (1 - \pi_k) \cdot \frac{y_k^2}{\pi_k^2} \cdot I_k^2$$

$$+ \sum_{U_Q} (1 - \pi_k) \cdot \frac{\frac{1}{b^2} \cdot (c \cdot y_k^2 + 2 \cdot (d - a \cdot b) \cdot y_k + e - a^2)}{\pi_k^2} \cdot I_k^2$$

applies. It follows that

$$E_P \left(E_Q \left(\sum \sum_s \frac{\Delta_{kl}}{\pi_{kl}} \cdot \frac{y_k'}{\pi_k} \cdot \frac{y_l'}{\pi_l} \,\middle|\, s \right) \right) = \sum_{U_D} (1 - \pi_k) \cdot \frac{y_k^2}{\pi_k}$$

$$+ \frac{1}{b^2} \cdot \sum_{U_Q} (1 - \pi_k) \cdot \frac{c \cdot y_k^2 + 2 \cdot (d - a \cdot b) \cdot y_k + e - a^2}{\pi_k}$$

$$+ \sum \sum_U \Delta_{kl} \cdot \frac{y_k}{\pi_k} \cdot \frac{y_l}{\pi_l} - \sum_U (1 - \pi_k) \cdot \frac{y_k^2}{\pi_k}$$

$$= \sum \sum_U \Delta_{kl} \cdot \frac{y_k}{\pi_k} \cdot \frac{y_l}{\pi_l} + \frac{1}{b^2} \cdot \left[(c - b^2) \cdot \sum_{U_Q} y_k^2 \cdot \left(\frac{1}{\pi_k} - 1 \right) \right.$$

$$\left. + 2 \cdot (d - a \cdot b) \cdot \sum_{U_Q} y_k \cdot \left(\frac{1}{\pi_k} - 1 \right) + (e - a^2) \cdot \sum_{U_Q} \left(\frac{1}{\pi_k} - 1 \right) \right].$$

Hence, for (6.29) to be unbiased for (6.28), terms have to be added to $\sum \sum_s \frac{\Delta_{kl}}{\pi_{kl}} \cdot \frac{y_k'}{\pi_k} \cdot \frac{y_l'}{\pi_l}$, for which the expectation over P and Q result in $\frac{1}{b^2} \cdot (c - b^2) \cdot \sum_{U_Q} y_k^2$, $\frac{1}{b^2} \cdot 2 \cdot (d - a \cdot b) \cdot \sum_{U_Q} y_k$, and $\frac{1}{b^2} \cdot (e - a^2) \cdot N_Q$, respectively. Therefore, with

$$E_P \left(E_Q \left(\frac{1}{c} \cdot \sum_{s_Q} ((y_k^i)^2 \cdot b^2 - 2 \cdot y_k^i \cdot (d - a \cdot b) - e + a^2) \cdot \frac{1}{\pi_k} \,\middle|\, s \right) \right)$$

$$= E_P \left(\sum_{s_Q} y_k^2 \cdot \frac{1}{\pi_k} \right) = \sum_{U_Q} y_k^2$$

for strategy L (compare to the specific derivation in the proof of Theorem 3c),

$$E_P \left(E_Q \left(\sum_{s_Q} y_k^i \cdot \frac{1}{\pi_k} \,\middle|\, s \right) \right) = \sum_{U_Q} y_k,$$

and

$$E_P \left(E_Q \left(\sum_{s_Q} \frac{1}{\pi_k} \,\middle|\, s \right) \right) = N_Q,$$

the proof is completed.

6.3.3 Accuracy and Privacy Protection

The measure of theoretical privacy protection considered in Sect. 6.2.4 for categorical variables makes no sense for quantitative variables. In this case, the privacy of a respondent may not be protected, although the given answer cannot be identified as the true one. How close the given answers z_k are compared with the true values y_k also has to be taken into account (for $k \in s$). Hence, a possible measure of disclosure is the expected squared difference between the answer z and the true value y, $E(z - y)^2$, as discussed by Zaizai et al. (2009). However, neither does this measure show that the respondents' privacy is well protected because $E(z - y)^2$ has no upper bound as mentioned by Diana and Perri (2011, p. 21), nor does a total disclosure of the respondents' true values of y mean that $E(z - y)^2 = 0$ has to apply. For instance, when the design Q is the multiplicative model $z_k = y_k \cdot u_k$ with $u_k = u$ being a non-zero constant and $p_{IV} = 1$, we can directly conclude from z_k to y_k for any u while $E(z - y)^2 = (u - 1)^2 \cdot (\sigma_y^2 + \mu_y^2)$ is zero only for $u = 1$.

An example of a sound measure of the level of disclosure offered by an RR design of our family Q with the desired properties is the squared correlation coefficient of z and y. It results in zero when the privacy is totally protected and one for a full disclosure. Hence,

$$\lambda^{(Q)} = 1 - |\rho_{zy}| \tag{6.30}$$

is a measure of the level of privacy protection offered by the RR questioning design Q with $0 \leq \lambda^{(Q)} \leq 1$. It is easy to show that $\lambda^{(Q)}$ is equal to zero only for a strategy Q, for which $p_I = 1$ applies. This is the direct questioning design with absolutely no privacy protection. The measure reaches its other extreme $\lambda^{(Q)} = 1$ for a questioning design Q with design probabilities $p_I, p_{III}, p_{IV} = 0$ and $p_{II} + p_V = 1$. For such design probabilities, the answer z_k of survey unit k contains absolutely no information on y_k. In such cases, the privacy of a respondent is completely protected. In any case, the efficiency of different members of the RR family Q should only be compared at the same objective level of privacy protection as measured, for instance, according to (6.30).

Chapter 7
A Unified Framework for Statistical Disclosure Control

7.1 Introduction

Among the various application fields of the pseudo-population concept in statistical surveys, the area of methods for statistical disclosure control (SDC) is exceptional in a certain sense. What is unique about SDC methods is that in contrast to almost all other procedures, they are not aimed at improving, but in deliberately reducing the quality of data, which are observed in statistical surveys of the official statistics or other institutions, in a controlled way.

These data are concerned with all kinds of fields such as employment, education, public health, or others. SDC is nothing else than a balancing act between compulsory data protection and the continuously increasing demand by the public for access to original data. As such data may contain sensitive information on natural or legal persons, such as information on poverty, addictive behavior, diseases, tax morality, or bank rating, the release of such microdata files is subject to the laws of data protection. Disclosure happens if the release of data allows an intruder to connect the surveyed information to certain population units. To avoid such personal loss of privacy protection, it might not suffice to just delete those variables that are directly linked to survey units, such as name, address, telephone numbers, or an artificial identification number such as the social security number. Some of the units might still be identifiable by the rest of their records if they own, for instance, a rare combination of different variables such as the municipality of the residence and a very large income. In such situations, other sensitive information also contained in a microdata file may be assignable to a certain individual. For this reason, methods of SDC that make the linking of sensitive information to individuals impossible have to be applied before data can be handed out to the public. With the increasing ability of data collection and storage—think of the keyword "big data"—the issue of privacy protection is becoming increasingly important (cf., for instance, Young and Ludloff 2011).

The ultimate goal of SDC can be formulated as "disseminating statistical information in such a way that individual information is sufficiently protected

© Springer International Publishing Switzerland 2015
A. Quatember, *Pseudo-Populations*, DOI 10.1007/978-3-319-11785-0_7

against *recognition* of the subjects to which it refers, while at the same time providing society with such as much information as possible under this restriction" (Willenborg and de Waal 1996, p. 2; emphasis in original). In concrete terms, this means that variables have to be manipulated in a manner that enhances privacy protection where it is still possible to estimate the unknown parameters of interest.

7.2   The CSI Family of Methods for SDC

In the relevant literature, several such methods are discussed (cf., for instance, Winkler 2004, or Matthews and Harel 2011) and implemented into statistical software (cf., for instance, Templ 2008). The simplest of these strategies artificially introduces missing values, instead of sensitive values of a variable y into the microdata file. This approach is called "suppression of data" (cf., for instance, Willenborg and de Waal 2001, p. 28). In this way, the sample s is artificially divided into a response set s_r of size n_r and a missing set s_m of size n_{s_m} ($s = s_r \cup s_m, s_r \cap s_m = \emptyset, n = n_{s_r} + n_{s_m}$). Clearly, after suppressing data, the estimation of parameters under study (such as population totals) from the available cases is as problematic as it would be in the presence of real nonresponse (see Sect. 3.1). Assuming the absence of untruthful answers, for instance, in the resulting decomposed HT estimator (3.2) of t, the sum $\sum_{s_m} y_k \cdot \frac{1}{\pi_k}$ simply cannot be calculated.

In this section, a general framework of a whole family of methods for SDC to mask a sensitive or identifying variable y in a data file, the CSI technique, is discussed. Within this framework, the masking process can be viewed a missing data problem. Consequently, this process allows implementing the wide field of data imputation techniques (see Sect. 3.3) and also strategies of randomized response (see Chap. 6) into the methods of SDC for microdata. Hence, it may serve as another example of the application of the pseudo-population concept.

The framework consists of four consecutive steps applied to a sample s of size n (cf. Quatember and Hausner 2013): In the first, the C-step, an additional variable y^C is created by simply cloning the original variable y ($y^C = y$). In the next, the S-step of the masking process, the idea of data suppression is applied locally or even globally to the y clone y^C, which generates a variable y^{CS} substituting y^C. For this variable y^{CS}, the records of the cloned item are set to missing for a single survey unit, a group of survey units, or all survey units. Hence, with respect to y^{CS}, the sample s is artificially divided into a response set s_r of size n_{s_r} with $y_k^{CS} = y_k \ \forall$ $k \in s_r$ and a set s_m of size n_{s_m}, where y_k^{CS} of the cloned and suppressed variable y^{CS} is missing $\forall \ k \in s_m$.

In the following I-step, a method to impute data y_k^i for these missings is applied. In contrast to a real nonresponse case, herein, not only information on available

Fig. 7.1 The four steps of the CSI method in statistical disclosure control

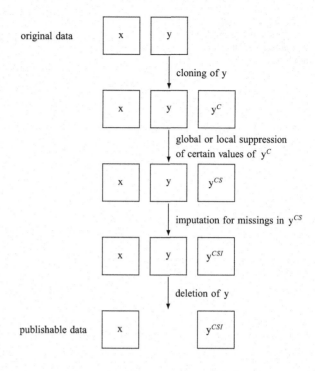

auxiliary variables **x**, but also on the original sensitive or identifying variable y can be used. In the end, the cloned and suppressed and imputed variable y^{CSI} has values

$$y_k^{CSI} = \begin{cases} y_k & \text{if unit } k \in s_r, \\ y_k^i & \text{otherwise.} \end{cases}$$

Note that the user of the data file should not be able to distinguish between true and imputed values of y^{CSI}. After the missing values y_k of y_k^{CS} in s_m have been replaced by imputed values y_k^i for variable y^{CSI}, in the concluding deletion step of the CSI process, the original variable y is completely deleted from the microdata file for all survey units. Henceforward, its masked substitute y^{CSI} has to serve as the basis for the estimation of the parameters under study in the publishable data file as far as this variable is concerned (see Fig. 7.1).

As an example, let the total t of variable y be the parameter of interest. With the original variable, the HT estimator t_{HT} according to (2.4) unbiasedly estimates t. When the CSI method described earlier is applied as the SDC technique to a without-replacement probability sample s resulting in a masked sample s_{CSI} with

Fig. 7.2 Generating a
pseudo-population within the
CSI process of statistical
disclosure control

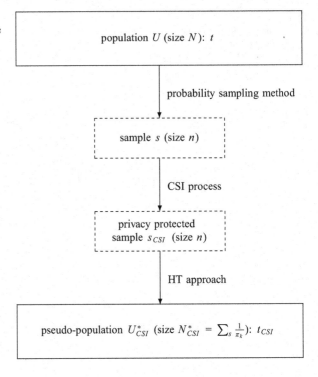

regard to the study variable y, the estimation of t can be done by the following estimator t_{CSI}, which corresponds to t_l in Eq. (3.4):

$$t_{\text{CSI}} = \sum_{s_{\text{CSI}}} y_k^{\text{CSI}} \cdot \frac{1}{\pi_k}. \tag{7.1}$$

The reasoning behind this is that the original population U of size N is estimated with regard to the estimation of total t of variable y by a pseudo-population U_{CSI}^* of size $N_{\text{CSI}}^* = \sum_s \frac{1}{\pi_k}$ (see Fig. 7.2). For this purpose, after cloning, suppression, imputation and deletion, the masked variable value y_k^{CSI} of each sample unit k in the sample s_{CSI} is replicated $\frac{1}{\pi_k}$ times ($k \in s$). Hence, in this context, estimator (7.1) can be presented as

$$t_{\text{CSI}} = \sum_{U_{\text{CSI}}^*} y_k^* \tag{7.2}$$

with the replications y^* of the y^{CSI}-values.

At this very point, the generation of the pseudo-population U_{CSI}^*, the balancing act of SDC between the two specific interests, mandatory data protection and the demand from researchers for access to original data, takes place. On the one hand, the quality of U_{CSI}^* as set-valued estimator of U with respect to the estimation of t

depends on the quality of the estimation of the second sum of Eq. (3.2), $\sum_{s_m} y_k \cdot \frac{1}{\pi_k}$, by $\sum_{s_m} y_k^i \cdot \frac{1}{\pi_k}$ using the imputed values of y^{CSI}. On the other hand, the degree of privacy protection offered by the CSI method to an element k of the sample depends on the prediction quality of y_k^i for y_k.

Therefore, the perfect technique of SDC with regard to the estimation of t would be a procedure that combines both interests. This could be done by a method that, on the one hand, does not allow to conclude from y_k^i in the privacy-protected sample s_{CSI} to the true value y_k in the original sample s, while on the other allows that the sum $\sum_{s_m} y_k^i \cdot \frac{1}{\pi_k}$ of the artificial missing set s_m estimates $\sum_{s_m} y_k \cdot \frac{1}{\pi_k}$ fairly well. The questions that have to be answered are:

- Which values of y^C should be suppressed or, in other words, which artificial nonresponse mechanism should be applied to the data set? and
- Which substitute values y_k^{CSI} should be imputed for these suppressed values ($k \in s_m$)?

Actually, the answers to these questions may differ from situation to situation. Possible methods to be applied at the I-step of the CSI process include the wide range of imputation techniques from the field of missing data. For example, if variable y observed in an SI sample s is sensitive as a whole, HD imputation as discussed in Sect. 3.3 can be applied in this context. The resulting procedure can be described in the following way: after the cloning of the original variable y in s, a subset s_m of n_{s_m} elements, a number which can be determined by the data supplier, is randomly set to missing with respect to the clone y^C. This mimics the MCAR item nonresponse mechanism (cf. Little and Rubin 2002, p. 12). Privacy protection with respect to study variable y increases with increasing proportion $\frac{n_{s_m}}{n}$ because the probability that y_k^{CSI} is equal to the true value y_k decreases. The ratio $\frac{n_{s_m}}{n}$ indirectly affects the composition of the pseudo-population U_{CSI}^* with respect to y^{CSI}. It can be called the artificial item nonresponse rate of this process. In the I-step of this method, the artificial missings are replaced by a random hot-deck imputation of values taken randomly with replacement from all n_{s_r} remaining values in the response set s_r of y^{CS}. This results in the masked variable y^{CSI}.

The more imputed replicates are included in U_{CSI}^*, the higher is the privacy protection and the lower is the efficiency of an estimator such as t_{CSI}. However, usual estimates of the variance $V(t_{\text{CSI}})$ may be too small unless the number of suppressed values is negligible because they do not account for imputation uncertainty. Here, multiple imputation may be helpful (see Sect. 3.3). Rubin (1993) proposed its use in the SDC context. When it is applied as an imputation method within the I-step of the CSI framework to replace all or a certain part of the y-clone y^C, the so-called "partially synthetic datasets" are generated (cf. Drechsler et al. 2008, p. 1007). All stochastic imputation methods may be used. For the estimation of the interesting parameters, the multiple imputation framework can be used (cf. Reiter 2003, p. 5f).

Of course, procedures developed originally in the SDC context can be used. For example, randomly interchanging the values of the sensitive variable of two different groups of the same size is called "data swapping" (cf. Dalenius and Reiss 1982).

This preserves the privacy of the units belonging to these groups. Hence, the real data remain in the data set, but some of them are exchanged in groups. Obviously, this does not affect the composition of pseudo-population U_{CSI}^* resulting from (7.1), in which each value y_k^{CSI} is replicated $\frac{1}{\pi_k}$ times, if s is a self-weighting sample. For such samples, this does not deteriorate the estimation quality of statistics for the distribution of y, but it protects the privacy of the respondents.

The term "micro-aggregation of data" (cf. Defays and Anwar 1998) refers to a strategy where sensitive values of a quantitative variable are—generally spoken— substituted by aggregates such as means, medians, modes, or some other measures. After the micro-aggregation, in particular in the case of mean imputation, simple univariate statistics of y such as its total may still be calculated in the pseudo-population U_{CSI}^*, which then consists of replicated true values and replicated aggregates. But, the variance of the variable of replicates $y^{\text{CSI}*}$ in U_{CSI}^* will certainly understate the true variance of y in U. When compared with the imputation of overall aggregates, the imputation of aggregates calculated within the same classes (of $\mathbf{x}$ and/or y) as the suppressed data will surely help in increasing the quality of the estimation of such parameters on the basis of the new variable y^{CSI}.

The addition of noise (cf. Fuller 1993) is another example of an SDC procedure belonging to the CSI family. Herein, in the imputation step, random errors are added to y to create the publishable variable y^{CSI}. This I-step can be seen as an application of stochastic regression imputation. The estimation of univariate parameters is without a problem, if the chosen suppression mechanism is MCAR. If it is only MAR, conditional stochastic regression imputation can be applied.

Furthermore, global recoding and top and/or bottom coding (see, for instance, Willenborg and de Waal 2001, p. 27f) also belong to the CSI family of SDC techniques. In these cases, the cloned data are globally or locally suppressed. The concluding I-step of the CSI process uses only the original variable y as auxiliary information and transforms its values, on the one hand, into large(r) intervals and, on the other hand, limits the extreme values of y to an upper and/or lower bound. This means a loss of information contained in the pseudo-population U_{CSI}^*, which may not very much affect the estimation of parameters when robust estimators with respect to outliers are calculated in the pseudo-population, such as the median.

Moreover, techniques of RR, originally presented as methods to reduce nonresponse and untruthful answering when sensitive questions such as harassment at work, domestic violence, or illegal employment are asked in a survey (see Sect. 6), can be incorporated in the CSI context. The central element of these methods is that survey units do not have to answer the sensitive question with certainty, but can choose the one to be answered randomly from two or more questions (or instructions). This does not enable the data collector to identify the question, on which the respondents have given their answers, although these answers still allow estimating the parameter under study. In this way, the idea is to reduce the individual's fear of an embarrassing "outing" to make sure that the responding person is willing to cooperate.

Warner (1971) was the first to indicate that these techniques are also applicable as methods of SDC applied during the data collection (cf. Warner 1971, p. 887).

In this case, the masking process is carried out by the respondents rather than the agency (cf. Quatember 2009, p. 144). In the I-step of the CSI strategy, the true value y_k of a cloned and globally suppressed variable y^{CS} is replaced by the value y_k^{CSI}, derived from the respondent's answer z_k on the randomly selected question.

In the SDC context, strategies, where a randomization mechanism is used as imputation algorithm after the data collection, are called "post randomization" techniques (see Gouweleeuw et al. 1998). In such cases, within the randomization device, the original variable value y_k may serve as auxiliary information. For RR technique Q presented in Sect. 6.3, this means that the randomization with respect to the variable y to be masked has to be done by the agency. This results in masked values

$$
y_k^{CSI} = \begin{cases}
y_k & \text{with probability } p_{\mathrm{I}}, \\
x_k & \text{with probability } p_{\mathrm{II}}, \\
y_k + u_k & \text{with probability } p_{\mathrm{III}}, \\
y_k \cdot v_k & \text{with probability } p_{\mathrm{IV}}, \\
w_k & \text{with probability } p_{\mathrm{V}}
\end{cases}
$$

$(k = 1, \ldots, n, \sum p_i = 1)$. For this purpose, all design probabilities and variables have to be chosen reasonably with respect to a trade-off between efficiency and data protection. For t_{CSI}, for instance, the application of the RR questioning design Q after the data collection results in the generation of a pseudo-population U_{CSI}^*, in which the sum of all values y_k^{CSI} estimates unbiasedly the total of y in U. The variance of t_{CSI} and an estimate for this variance can be calculated with (6.21) and (6.25), respectively. These formulae contain terms that correspond to the loss of efficiency caused by the performed data protection. By choosing the design probabilities and parameters of Q, both the efficiency of the mean estimation and the level of privacy protection calculated according to (6.30) can be controlled and balanced against each other.

Also, as described in Sects. 6.2.3 and 6.3.2 for categorical and quantitative variables in the RR context, randomization does not have to be applied to all survey units. In the field of SDC, this means that the sample s is not completely set to missing with regard to variable y^C $(s_m \subset s)$.

After all, there should be no doubt about it that it is absolutely necessary with respect to the quality of the results regarding efficiency and data protection to use the experience of practitioners such as members of national statistical agencies. With their knowledge, the questions previously asked regarding the CSI method to be applied could be answered satisfactorily. Also multivariate relationships between surveyed variables may be maintained in most cases, when an efficient masking process can simultaneously be applied to different variables.

The application of a member of the CSI family of techniques to create a publishable microdata file by masking a sensitive variable y has two unavoidable impacts. On the one hand, information is lost, and on the other hand, the privacy of respondents is better protected than without. With regard to the first impact, the loss of data quality can be measured, for instance, by the amount of increased bias and mean square error of estimators such as t_{CSI} (7.1), when compared with the estimators applied to the original data. This loss of efficiency of estimators can be interpreted as the price that has to be paid by the data users for the increased privacy protection of the survey units.

The privacy protection corresponding to a certain CSI method can be quantified, for instance, by a simple measure such as the relative size $\frac{n_{s_m}}{n}$ of the missing set s_m created at the S-step of the process. More sophisticated methods relate the publishable variable y^{CSI} to the sensitive variable y. For binary variables, measures published, for instance, by Quatember (2009) or in Sect. 7.2 of Chaudhuri and Christofides (2013) use the conditional probabilities of certain answers given the possession or nonpossession of the sensitive attribute. For categorical variables, such measures as $\lambda_h^{(R)}$ (6.16) presented in Sect. 6.2.4 can be applied. For quantitative variables (see Sect. 6.3.3), these include the squared correlation coefficient between variables y^{CSI} and y (cf. Diana et al. 2013, p. 20f), and the mean value of the squared differences of y^{CSI} and y over all survey units in the cloned, suppressed and imputed sample before the original variable y is deleted (cf. Zaizai et al. 2009). A squared correlation of one and a sum of squared differences resulting in zero correspond to the complete absence of data protection in the microdata file with regard to the sensitive variable y.

These measures of data protection can be calculated by the data supplier after the I-step of the process before the microdata file is ready to be published. But that is another story that has nothing to do with the main topic of the book—the generation of pseudo-populations.

References

Alfons, A., Kraft, S., Templ, M., & Filzmoser, P. (2011a). Simulation of close-to-reality population data for household surveys with application to EU-SILC. *Statistical Methods and Applications, 20*(3), 383–407.

Alfons, A., Templ, M., Burgard, J., Filzmoser, P., Hulliger, B., Kolb, J. -P., et al. (2011b). *The AMELI simulation study*. Research Project Report WP6 (D6.1), FP7-SSH-2007-217322 AMELI. Available on [September 10, 2015]: https://www.uni-trier.de/index.php?id=24676.

Antal, E., & Tillé, Y. (2011). A direct bootstrap method for complex sampling designs from a finite population. *Journal of the American Statistical Association, 106*(494), 534–543.

Ardilly, P., & Tillé, Y. (2006). *Sampling methods: Exercises and solutions*. New York: Springer.

Baker, R., Brick, J. M., Bates, N. A., Battaglia, M., Couper, M. P., Dever, J. A., et al. (Eds.). (2013). Report of the AAPOR task force on non-probability sampling. American Association of Publich Opinion Research, AAPOR Publishing. Available on [September 10, 2015]: http://www.aapor.org/AM/Template.cfm?Section=Reports1&Template=/CM/ContentDisplay.cfm&ContentID=5963.

Bar-Lev, S. K., Bobovitch, E., & Boukai, B. (2004). A note on randomized response models for quantitative data. *Metrika, 60*, 255–260.

Barbiero, A., & Mecatti, F. (2010). Bootstrap algorithms for variance estimation in πPS sampling. In P. Mantovan, P. Secchi (Eds.), *Complex data modeling and computationally intensive statistical methods* (pp. 57–69). Milan: Springer.

Bauer, A. (2011). Genauigkeitsanalysen zu den Österreich-Ergebnissen der PISA-Studie 2009. Master Thesis, Johannes Kepler University Linz.

Beatty, P., & Herrmann, D. (1995). A framework for evaluating "Don't Know" responses in surveys. *Proceedings of the Survey Research Methods Section of the American Statistical Association*, 1005–1010.

Beatty, P., & Herrmann, D. (2002). To answer or not to answer: Decision processes related to survey item nonresponse. In R. M. Groves, D. A. Dillman, J. L. Eltinge, & R. J. A. Little (Eds.), *Survey nonresponse* (pp. 71–86). New York: Wiley.

Bellhouse, D. R. (1988). A brief history of random sampling methods. In P. R. Krishnaiah, & C. R. Rao (Eds.), *Handbook of statistics* (Vol. 6, pp. 1–14). Amsterdam: Elsevier.

Bethlehem, J. (2002). Weighting nonresponse adjustments based on auxiliary information. In R. M. Groves, D. A. Dillman, J. L. Eltinge, & R. J. A. Little (Eds.), *Survey nonresponse* (pp. 275–287). New York: Wiley.

Bethlehem, J. (2009). The rise of survey sampling (Discussion paper (09015)). The Hague/Heerlen: Statistics Netherlands. Available on [September 10, 2015]: http://www.cbs.nl/NR/rdonlyres/BD480FBC-24CF-42FA-9A0D-BBECD4F53090/0/200915x10pub.pdf.

© Springer International Publishing Switzerland 2015
A. Quatember, *Pseudo-Populations*, DOI 10.1007/978-3-319-11785-0

Bhargava, M., & Singh, R. (2000). A modified randomization device for Warner's model. *Statistica, LX*, 315–321.

Bickel, P. J., & Freedman, D. A. (1984). Asymptotic normality and the bootstrap in stratified sampling. *The Annals of Statistics, 12*(2), 470–482.

Boeije, H., & Lensvelt-Mulders, G. (2002). Honest by chance: A qualitative study to clarify respondents' (non-)compliance with computer-assisted randomized response. *Bulletin de Methodologie Sociologique, 75*, 24–39.

Booth, J. G., Butler, R. W., & Hall, P. (1994). Bootstrap methods for finite populations. *Journal of the American Statistical Association, 89*(428), 1282–1289.

Boruch, R. F. (1971). Assuring confidentiality of responses in social research: A note on strategies. *The American Sociologist, 6*, 308–311.

Botman, S. L., & Thornberry, O. T. (1992). Survey design features correlates of nonresponse. *Proceedings of the Survey Research Methods Section of the American Statistical Association*, 309–314.

Bourke, P. D. (1984). Estimation of proportions using symmetric randomized response designs. *Psychological Bulletin, 96*(1), 166–172.

Bourke, P. D., & Dalenius, T. (1976). Some new ideas in the realm of randomized inquiries. *International Statistical Review, 44*(2), 219–221.

Bourke, P. D., & Moran, M. M. (1988). Estimating proportions from randomized response data using the EM algorithm. *Journal of the American Statistical Association, 83*(404), 964–968.

Buckland, S. T. (1984). Monte Carlo confidence intervals. *Biometrics, 40*, 811–817.

Casella, G., & Berger, R. L. (2002). *Statistical inference* (2nd ed.). Cengage Learning, Brooks/Cole.

Chang, H. -J., Wang, C. -L., & Huang, K. -C. (2004). On estimating the proportion of a qualitative sensitive character using randomized response sampling. *Quality & Quantity, 38*, 675–680.

Chao, M. -T., & Lo, S. -H. (1985). A bootstrap method for finite populations. *Sankhya, Series A, 47*, 399–405.

Chao, M. -T., & Lo, S. -H. (1994). Maximum likelihood summary and the bootstrap method in structured finite populations. *Statistica Sinica, 4*, 389–406.

Chapman, D. G. (1951). Some properties of the hypergeometric distribution with applications to zoological censuses. *University of California Publications in Statistics, 1*, 131–160. Available on [September 10, 2015]: http://babel.hathitrust.org/cgi/pt?id=wu.89045844248#view=1up; seq=22.

Chaudhuri, A. (2001). Using randomized response from a complex survey to estimate a sensitive proportion in a dichotomous finite population. *Journal of Statistical Planning and Inference, 94*, 37–42.

Chaudhuri, A. (2011). *Randomized response and indirect questioning techniques in surveys.* Boca Raton: CRC Press.

Chaudhuri, A., & Christofides, T. C. (2013). *Indirect questioning in sample surveys.* Heidelberg: Springer.

Christofides, T. C. (2003). A generalized randomized response technique. *Metrika, 57*, 195–200.

Church, A. H. (1993). Estimating the effect of incentives on mail survey response rates: A meta-analysis. *The Public Opinion Quarterly, 57*(1), 62–79.

Cochran, W. G. (1977). *Sampling techniques* (3rd ed.). New York: Wiley.

Dalenius, T., & Reiss, S. P. (1982). Data-swapping: A technique for disclosure control. *Journal of Statistical Planning and Inference, 6*, 73–85.

Defays, D., & Anwar, M. N. (1998). Masking microdata using micro-aggregation. *Journal of Official Statistics, 14*(4), 449–461.

Dempster, A. P., Laird, N. M., & Rubin, D. B. (1977). Maximum likelihood from incomplete data via the EM algorithm. *Journal of the Royal Statistical Society, Series B, 39*(1), 1–38.

Deville, J.-C., & Tillé, Y. (2004). Efficient balanced sampling: The cube method. *Biometrika, 91*(4), 893–912.

Diana, G., Giordan, M., & Perri, P. F. (2013). Randomized response surveys: A note on some privacy protection measures. *Model Assisted Statistics and Applications, 8*, 19–28.

Diana, G., & Perri, P. F. (2009). Estimating a sensitive proportion through randomized response procedures based on auxiliary information. *Statistical Papers, 50*, 661–672.

Diana, G., & Perri, P. F. (2011). A class of estimators for quantitative sensitive data. *Statistical Papers, 52*, 633–650.

Diekmann, A. (2012). Making use of "Benford's Law" for the randomized response technique. *Sociological Methods & Research, 41*(2), 325–334.

Dillman, D. A. (1978). *Mail and telephone surveys: The total design method.* New York: Wiley InterScience.

Dillman, D. A. (2000). *Mail and Internet surveys.* New York: Wiley.

Dillman, D. A., Singer, E., Clark, J. R., & Treat, J. B. (1996). Effects of benefits appeals, mandatory appeals, and variations in statements of confidentiality on completion rates for census questionnaires. *The Public Opinion Quarterly, 60*(3), 376–389.

Drechsler, J., Bender, S. & Rässler, S. (2008). Comparing fully and partially synthetic datasets for statistical disclosure control in the German IAB establishment panel. *Transactions on Data Privacy, 1*, 1002–1050.

Efron, B. (1979). Bootstrap methods: Another look at the jackknife. *Annals of Statistics, 7*, 1–26.

Efron, B. (1981). Censored data and the bootstrap. *Journal of the American Statistical Association, 76*(374), 312–319.

Eichhorn, B. H., & Hayre, L. S. (1983). Scrambled randomized response methods for obtaining sensitive quantitative data. *Journal of Statistical Planning and Inference, 7*, 307–316.

Eurostat. (2012). *Labour force survey in the EU, candidate and EFTA countries, 2012 Edition.* Luxembourg: Publications Office of the European Union. Available on [September 10, 2015]: http://epp.eurostat.ec.europa.eu/portal/page/portal/employment_unemployment_lfs/publications/methods.

Fay, R. E. (1989). Theory and application of replicate weighting for variance calculations. *Proceedings of the Survey Research Methods Section of the American Statistical Association*, 212–217.

Fienberg, S. E. (1992). Bibliography on capture-recapture modelling with application to census undercount adjustment. *Survey Methodology, 18*(1), 143–154.

Fuller, W. A. (1993). Masking procedures for microdata disclosure limitation. *Journal of Official Statistics, 9*(2), 383–406.

Gabler, S., & Häder, S. (2009). Die Kombination von Mobilfunk- und Festnetzstichproben in Deutschland. In M. Weichbold, J. Bacher, & C. Wolf (Eds.), *Umfrageforschung* (pp. 499–512). Wiesbaden: VS Verlag für Sozialwissenschaften.

Gabler, S., & Quatember, A. (2012). Das Problem mit der Repräsentativität von Stichprobenerhebungen. In vsms (Verband Schweizer Markt-und Sozialforschung) (Ed.), *Jahrbuch 2012* (pp. 17–19). Zürich: vsms.

Gabler, S., & Quatember, A. (2013). Repräsentativität von Subgruppen bei geschichteten Zufallsstichproben. *AStA Wirtschafts- und Sozialstatistisches Archiv, 7*, 105–119.

Gjestvang, C. R., & Singh, S. (2007). Forced quantitative randomized response model: A new device. *Metrika, 66*, 243–257.

Gouweleeuw, J. M., Kooiman, P., Willenborg, L. C. R. J., & de Wolf, P. -P. (1998). Post randomisation for statistical disclosure control: Theory and implementation. *Journal of Official Statistics, 14*(4), 463–478.

Greenberg, B. G., Abul-Ela, A. -L. A., Simmons, W. R., & Horvitz, D. G. (1969). The unrelated question randomized response model: Theoretical framework. *Journal of the American Statistical Association, 64*(326), 520–539.

Greenberg, B. G., Kuebler, R. R. Jr., Abernathy, J. R., & Horvitz, D. G. (1971). Application of the randomized response technique in obtaining quantitative data. *Journal of the American Statistical Association, 66*(334), 243–250.

Groenitz, H. (2014). A new privacy-protecting survey design for multichotomous sensitive variables. *Metrika, 77*, 211–224.

Gross, S. (1980). Median estimation in sample surveys. *Proceedings of the Survey Research Methods Section of the American Statistical Association*, 181–184.

Groves, R. M., Fowler, F. J., Couper, M. P., Lepkowski, J. M., Singer, E., & Tourangeau, R. (2004). *Survey methodology.* Hoboken: Wiley.

Groves, R. M., & Heeringa, S. G. (2006). Responsive design for household surveys: Tools for actively controlling survey errors and costs. *Journal of the Royal Statistical Society, Series A, 169*(3), 439–457.

Guerriero, M., & Sandri, M. F. (2007). A note on the comparison of some randomized response procedures. *Journal of Statistical Planning and Inference, 137*, 2184–2190.

Hansen, M. H., & Hurwitz, W. N. (1943). On the theory of sampling from finite populations. *Annals of Mathematical Statistics, 14*, 333–362.

Heberlein, T. A., & Baumgartner, R. (1978). Factors affecting response rates to mailed questionnaires: A quantitative analysis of the published literature. *American Sociological Review, 43*(4), 447–462.

Holbrook, A. L., & Krosnick, J. A. (2010). Measuring voter turnout by using the randomized response technique. *The Public Opinion Quarterly, 74*(2), 328–343.

Holmberg, A. (1998). A bootstrap approach to probability proportional-to-size sampling. *Proceedings of the Survey Research Methods Section of the American Statistical Association,* 378–383.

Holt, A., & Elliot, D. (1991). Methods of weighting for unit-nonresponse. *The Statistician, 40*(3), 333–342.

Horvitz, D. G., Shah, B. V., & Simmons, W. R. (1967). The unrelated question randomized response model. *Proceedings of the Social Statistics Section of the American Statistical Association,* 65–72.

Horvitz, D. G., & Thompson, D. J. (1952). A generalization of sampling without replacement from a finite universe. *Journal of the American Statistical Association, 47*, 663–685.

International Working Group for Disease Monitoring and Forecasting. (1995a). Capture-recapture and multiple-record systems estimation I: History and theoretical development. *American Journal of Epidemiology, 142*(10), 1047–1058.

International Working Group for Disease Monitoring and Forecasting. (1995b). Capture-recapture and multiple-record systems estimation II: Applications in human diseases. *American Journal of Epidemiology, 142*(10), 1059–1068.

James, J. M., & Bolstein, R. (1990). The effect of monetary incentives and follow-up mailings on the response rate and response quality in mail surveys. *The Public Opinion Quarterly, 54*(3), 346–361.

Judkins, D. R. (1990). Fay's method for variance estimation. *Journal of Official Statistics, 6*(3), 223–239.

Kauermann, G., & Küchenhoff, H. (2011). *Stichproben.* Heidelberg: Springer.

Kreuter, F. (2008). Interviewer effects. In P. J. Lavrakas (Ed.), *Encyclopedia of survey research methods* (pp. 369–371). Los Angeles: SAGE Publications.

Kruskal, W., & Mosteller, F. (1980). Representative sampling, IV: The history of the concept in statistics, 1895–1939. *International Statistical Review, 48*, 169–195.

Kuk, A. Y. C. (1989). Double bootstrap estimation of variance under systematic sampling with probability proportional to Size. *Journal of Statistical Computation and Simulation, 31*, 73–82.

Lahiri, P. (2003). On the impact of bootstrap in survey sampling and small-area estimation. *Statistical Science, 18*(2), 199–210.

Lensvelt-Mulders, G. J. L. M., Hox, J. J., van der Heijden, P. G. M., & Maas, C. J. M. (2005). Meta-analysis of randomized response research. *Sociological Methods & Research, 33*(3), 319–348.

Little, R. J. A., & Rubin, D. B. (2002). *Statistical analysis with missing data.* Hoboken: Wiley.

Liu, P. T., & Chow, L. P. (1976). A new discrete quantitative randomized response model. *Journal of the American Statistical Association, 71*(353), 30–31.

Lohr, S. L. (2010). *Sampling: Design and analysis* (2nd ed.). Boston: Brooks/Cole.

Lumley, T. (2010). *Complex surveys - a guide to analysis using R.* Hoboken: John Wiley.

Mangat, N. S. (1992). Two stage randomized response sampling procedure using unrelated question. *Journal of the Indian Society of Agricultural Statistics, 44*, 82–87.

Mangat, N. S., & Singh, R. (1990). An alternative randomized response procedure. *Biometrika, 77*, 439–442.

Mangat, N. S., & Singh, R. (1991). An alternative approach to randomized response survey. *Statistica, 3*, 327–332.

Mangat, N. S., Singh, S., & Singh, R. (1993). On the use of a modified randomization device in randomized response inquiries. *Metron, 51*, 211–216.

Mangat, N. S., Singh, S., & Singh, R. (1995). On use of a modified randomization device in Warner's model. *Journal of Indian Society of Statistics & Operations Research, 16*, 65–69.

Matthews, G. J., & Harel, O. (2011). Data confidentiality: A review of methods for statistical disclosure limitation and methods for assessing privacy. *Statistics Surveys, 5*, 1–29.

McCarthy, P. J., & Snowden, C. B. (1985). The bootstrap and finite population sampling. *Vital and Health Statistics*, Series 2, No. 95, Public Health Service, Washington, U.S. Government Printing Office.

Mislevy, R. J. (1991). Randomization-based inference about latent variables from complex samples. *Psychometrika, 56*(2), 177–196.

Münnich, R., Magg, K., Sostra, K., Schmidt, K., & Wiegert, R. (2004). *Variance estimation for small area estimates*. Research Project Report WP10 (D10.1 and 10.2), IST-2000-26057 DACSEIS. Available on [September 10, 2015]: http://www.uni-trier.de/index.php?id=29730.

Münnich, R., & Schürle, J. (2003). *On the simulation of complex universes in the case of applying the German microcensus*. DACSEIS Research Paper Series 4. Available on [September 10, 2015]: https://www.uni-trier.de/fileadmin/fb4/projekte/SurveyStatisticsNet/DRPS4.pdf.

Münnich, R., Schürle, J., Bihler, W., Boonstra, H. -J., Eckmair, D., Haslinger, A., et al. (2003). *Monte-Carlo simulation study of European surveys*. Research Project Report WP3 (D3.1 and 3.2), IST-2000-26057 DACSEIS. Available on [September 10, 2015]: http://www.uni-trier.de/index.php?id=29730.

Newcomb, S. (1881). Note on the frequency of use of the different digits in natural numbers. *American Journal of Mathematics, 4*, 39–40.

OECD. (Ed.). (2012). *PISA 2009 Technical Report*. PISA, OECD Publishing. Available on [September 10, 2015]: http://dx.doi.org/10.1787/9789264167872-en.

Peeters, C. F. W., Lensvelt-Mulders, G. J. L. M., & Lasthuizen, K. (2010). A note on a simple and practical randomized response framework for eliciting sensitive dichotomous and quantitative information. *Sociological Methods & Research, 39*(2), 283–296.

Perri, P. F. (2008). Modified randomized devices for Simmons' model. *Model Assisted Statistics and Applications, 3*, 233–239.

Pollock, K. H., & Bek, Y. (1976). A comparison of three randomized response models for quantitative data. *Journal of the American Statistical Association, 71*(356), 884–886.

Poole, W. K. (1974). Estimation of the distribution function of a contiunous type random variable through randomized response. *Journal of the American Statistical Association, 69*(348), 1002–1005.

Quatember, A. (1996a). Das Problem mit dem Begriff Repräsentativität. *Allgemeines Statistisches Archiv, 80*(2), 236–241.

Quatember, A. (1996b). *Das Quotenverfahren*. Linz: Universitätsverlag Rudolf Trauner.

Quatember, A. (2001a). Das Jahrhundert der Stichproben. *Austrian Journal of Statistics, 30*(1), 45–60.

Quatember, A. (2001b). *Die Quotenverfahren. Stichprobentheorie und -praxis*. Aachen: Shaker Verlag.

Quatember, A. (2009). A standardization of randomized response strategies. *Survey Methodology, 35*(2), 143–152.

Quatember, A. (2012). An extension of the standardized randomized response technique to a multi-stage setup. *Statistical Methods & Applications, 21*(4), 475–484.

Quatember, A. (2014a). A randomized response design for a polychotomous sensitive population and its application to opinion polls. *Model Assisted Statistics and Applications, 9*, 11–23.

Quatember, A. (2014b). The finite population bootstrap - from the maximum likelihood to the Horvitz-Thompson approach. *Austrian Journal of Statistics, 43*(2), 93–102.

Quatember, A., & Bauer, A. (2012). Genauigkeitsanalysen zu den Österreich-Ergebnissen der PISA-Studie 2009. In F. Eder (Ed.), *PISA 2009 - Nationale Zusatzanalysen* (pp. 534–550). Münster: Waxmann Verlag.

Quatember, A., & Hausner, M. C. (2013). A family of methods for statistical disclosure control. *Journal of Applied Statistics, 40*(2), 337–346.

Ranalli, M. G., & Mecatti, F. (2012). Comparing recent approaches for bootstrapping sample survey data: A first step towards a unified approach. *Proceedings of the Survey Research Methods Section of the American Statistical Association*, 4088–4099.

Rao, J. N. K. (2003). *Small area estimation*. Hoboken: Wiley.

Rao, J. N. K., & Thomas, D. R. (1988). The analysis of cross-classified categorical data from complex surveys. *Sociological Methodology, 18*, 213–269.

Rao, J. N. K., & Wu, C. F. J. (1988). Resampling inference with complex survey data. *Journal of the American Statistical Association, 83*(401), 231–241.

Reiter, J. P. (2003). Inference for partially synthetic, public use microdata sets. *Survey Methodology, 29*(2), 181–188.

Rosén, B. (1997). On sampling with probability proportional to size. *Journal of Statistical Planning and Inference, 62*, 159–191.

Rubin, D. B. (1987). *Multiple imputation for nonresponse in surveys*. New York: Wiley.

Rubin, D. B. (1993). Discussion: Statistical disclosure limitation. *Journal of Official Statistics, 9*(2), 461–468.

Ryu, J. -B., Kim, J. -M., Heo, T. -Y., & Park, C. G. (2005). On stratified randomized response sampling. *Model Assisted Statistics and Applications, 1*(1), 31–36.

Seber, G. A. F. (1970). The effect of trap response on tag recapture estimates. *Biometrics, 26*, 13–22.

Sekar, C., & Deming, E. W. (1949). On a method of estimating birth and death rates and extent of registration. *Journal of the American Statistical Association, 44*, 101–115.

Sen, A. R. (1953). On the estimate of the variance in sampling with varying probabilities. *Journal of the Indian Society of Agricultural Statistics, 5*, 119–127.

Shao, J., & Sitter, R. R. (1996). Bootstrap for imputed survey data. *Journal of the American Statistical Association, 91*(435), 1278–1288.

Shao, J., & Tu, D. (1995). *The jackknife and bootstrap*. New York: Springer.

Singer, E. (2002). The use of incentives to reduce nonresponse in household surveys. In R. M. Groves, D. A. Dillman, J. L. Eltinge, & R. J. A. Little (Eds.), *Survey nonresponse* (pp. 163–178). New York: Wiley.

Singer, E., Mathiowetz, N. A., & Couper, M. P. (1993). The impact of privacy and confidentiality concerns on survey participation: The case of the 1990 U.S. census. *The Public Opinion Quarterly, 57*(4), 465–482.

Singer, E., von Hoewyk, J., Gebler, N., Raghunathan, T., & McGonagle, K. (1999). The effect of incentives on response rates in interviewer-mediated surveys. *Journal of Official Statistics, 15*(2), 217–230.

Singer, E., van Hoewyk, J., & Maher, M. P. (2000). Experiments with incentives in telephone surveys. *The Public Opinion Quarterly, 64*(2), 171–188.

Singer, E., van Hoewyk, J., & Neugebauer, R. J. (2003). Attitudes and behavior: The impact of privacy and confidentiality concerns on participation in the 2000 census. *The Public Opinion Quarterly, 67*(3), 368–384.

Singer, E., von Thurn, D. R., & Miller, E. R. (1995). Confidentiality assurances and response: A quantitative review of the experimental literature. *The Public Opinion Quarterly, 59*(1), 66–77.

Singh, S., Horn, S., Singh, R., & Mangat, N. S. (2003). On the use of modified randomization device for estimating the prevalence of a sensitive attribute. *Statistics in Transition, 6*, 515–522.

Singh, S., & Sedory, S. A. (2011). Cramer-Rao lower bound of variance in randomized response sampling. *Sociological Methods & Research, 40*(3), 536–546.

Singh, S., Singh, R., Mangat, N. S., & Tracy, D. S. (1994). An alternative device for randomized responses. *Statistica, 54*, 233–243.

Singh, R., Singh, S., Mangat, N. S., & Tracy, D. S. (1995). An improved two stage randomized response strategy. *Statistical Papers, 36,* 265–271.

Sitter, R. R. (1992a). A resampling procedure for complex survey data. *Journal of the American Statistical Association, 87*(419), 755–765.

Sitter, R. R. (1992b). Comparing three bootstrap methods for survey data. *The Canadian Journal of Statistics, 20*(2), 135–154.

Särndal, C. -E., Swensson, B., & Wretman, J. (1992). *Model assisted survey sampling.* New York: Springer.

Tan, M. T., Tian, G. L., & Tang, M. L. (2009). Sample surveys with sensitive questions: A nonrandomized response approach. *The American Statistician, 63,* 9–16.

Templ, M. (2008). Statistical disclosure control for microdata using the R-package sdcMicro. *Transactions on Data Privacy, 1,* 67–85

Templ, M., Münnich, R., Alfons, A., Filzmoser, P., Hulliger, B., Kolb, J. -P., et al. (2011). *Synthetic data generation of SILC data.* Research Project Report WP6 (D6.2), FP7-SSH-2007-217322 AMELI. Available on [September 10, 2015]: https://www.uni-trier.de/index.php?id=24676.

Thomas, D. R., Singh, A. C., & Roberts, G. R. (1995). Independence tests for two-way tables under cluster sampling. *Proceedings of the Survey Research Methods Section of the American Statistical Association,* 885–890.

Tourangeau, R., & Smith, T. W. (1996). Asking sensitive questions: The impact of data collection mode, question format, and question context. *The Public Opinion Quarterly, 60*(2), 275–304.

Traugott, M. W., Groves, R. M., & Lepkowski, J. M. (1987). Using dual frame designs to reduce nonresponse in telephone surveys. *The Public Opinion Quarterly, 51*(4), 522–539.

van den Hout, A., & van der Heijden, P. G. M. (2002). Randomized response, statistical disclosure control and misclassification: A review. *International Statistical Review, 70*(2), 269–288.

Warner, S. L. (1965). Randomized response: A survey technique for eliminating evasive answer bias. *Journal of the American Statistical Association, 60,* 63–69.

Warner, S. L. (1971). The linear randomized response model. *Journal of the American Statistical Association, 66,* 884–888.

Warner, S. L. (1976). Optimal randomized response models. *International Statistical Review, 44*(2), 205–212.

Warner, S. L. (1986). The omitted digit randomized response model for telephone applications. *Proceedings of the Social Survey Research Methods Section of the American Statistical Association,* 441–443.

Willenborg, L., & de Waal, T. (1996). *Statistical disclosure control in practice.* New York: Springer.

Willenborg, L., & de Waal, T. (2001). *Elements of statistical disclosure control.* New York: Springer.

Winkler, W. E. (2004). *Masking and re-identification methods for public-use microdata: Overview and research problems.* Research Report Series of the Statistical Research Division of the U.S. Bureau of the Census. 2004–2006.

Wolter, K. M. (2007). *Introduction to variance estimation.* New York: Springer.

Yates, F., & Grundy, P. M. (1953). Selection without replacement from within strata with probability proportional to size. *Journal of the Royal Statistical Society, Series B, 15,* 235–261.

Young, T., & Ludloff, M. E. (2011). *Privacy and big data.* Sebastopol: O'Reilly Media.

Zaizai, Y., Jingyu, W., & Junfeng, L. (2009). An efficiency and protection degree-based comparison among the quantitative randomized response strategies. *Communications in Statistics: Theory and Methods, 38*(3), 400–408.

Subject Index

© Springer International Publishing Switzerland 2015
A. Quatember, *Pseudo-Populations*, DOI 10.1007/978-3-319-11785-0

Printed in the United States
By Bookmasters